ROCKS AND MINERALS

ROCKS AND MINERALS
A PHOTOGRAPHIC FIELD GUIDE

Chris and Helen Pellant

BLOOMSBURY
LONDON · NEW DELHI · NEW YORK · SYDNEY

First published in 2014 by Bloomsbury Publishing Plc,
50 Bedford Square, London WC1B 3DP

www.bloomsbury.com

ISBN (print) 978-1-47290-993-0

Bloomsbury Publishing, London, New Delhi, New York and Sydney

Bloomsbury is a trademark of Bloomsbury Publishing Plc

A CIP catalogue record for this book is available from the British Library

Publisher: Nigel Redman
Project editor: Alice Ward
Design: Susan McIntyre

Printed in China by Toppan Leefung Printing Co Ltd.

This book is produced using paper that is made from wood grown in
managed sustainable forests. It is natural, renewable and recyclable. The
logging and manufacturing processes conform to the environmental
regulation of the country of origin.

10 9 8 7 6 5 4 3 2 1

MIX
Paper from
responsible sources
FSC® C104723

FRONT COVER: *Part of Uluru (Auscape/UIG Gettyimages), Amethyst (Shutterstock)*
BACK COVER: *Gypsum, Fluorite and Pyrite, Agate, Larvikite (Chris and Helen Pellant), Sandstone formation (Shutterstock)*

CONTENTS

ABOVE: *A hot spring at Rotorua, North Island, New Zealand, deposits minerals around its opening.*

LEFT: *Citrine, an orange-brown variety of quartz, showing hexagonal pyramidal shapes and vitreous lustre. Specimen from Brazil.*

BELOW: *Weathered pinnacles of basalt at The Storr, Skye, Scotland.*

INTRODUCTION

Rocks and minerals are the basis for the structure of the Earth's crust. Minerals are composed of atoms of single elements or a number of different elements. They can have perfect crystalline form and amazingly rich colours; some are prized as gemstones; many are the source of economically useful metals and other raw materials. New minerals are still being discovered, especially in industrial slag and even on shipwrecks. Rocks are made of minerals. Many, especially crystalline igneous rocks and marble, are frequently used decoratively, and certain rocks such as limestone, coal and ironstone have great economic uses. However, the Earth has been scarred by the extraction and use of these raw materials; watercourses and soils have been polluted and the atmosphere changed.

Our own interest in rocks and minerals began many years ago and is related to our delight in, and study of, the natural world. Minerals have always held a fascination and our collection has grown over the years. Many of the illustrations in the book are of our own specimens.

Making a collection of rock and mineral specimens may require fieldwork. Good locations can be researched using local geology field guides, geological maps and the internet. It is essential to obtain permission to gain access to private land and always to take great care when near cliffs or quarry faces. When breaking rocks, a geological hammer should always be used and eye protection worn. Specimens should be carefully cleaned and curated, with a card label giving location, date and other details. Fine or rare specimens can be bought from dealers.

This book is designed to help the reader identify and understand some of the more common rocks and minerals. The introductory sections outline basic formation and methods of classification and identification; the main sections, with illustrations, outline each of the rocks and minerals. The items selected and illustrated are those which may be found or obtained with relative ease.

ABOVE: Zigzag folding of alternating sandstone and shale at Millook Haven, Cornwall, England. *BELOW:* Granite exposure, Guernsey, Channel Islands, showing characteristic jointed structure emphasised by weathering.

ROCKS

Rocks are the basic materials of which our planet and many others in the solar system are composed. They are aggregates of minerals, commonly a few, sometimes only one or two. Dark crystalline basalt and loose sand both fit this definition. The characteristic features of any rock, and hence the means by which it can be identified and named, are a direct result of how it formed. Rocks are thus divided into three broad groups, the igneous, metamorphic and sedimentary rocks. Igneous rocks result from the cooling and consolidation of magma or lava; metamorphic rocks are those which have been changed from their original state by heat and/or pressure; sedimentary rocks are usually deposited in strata (layers) and may result from the weathering and erosion of pre-existing rocks.

Rocks form in a cycle. The first to develop, the primary rocks, are the igneous rocks. Magma (molten rock underground) and lava (molten rock on the surface) come from some depth in the crust or upper mantle of the Earth. Once magma or lava has cooled, and igneous rocks have formed, weathering and erosion can break them down into their constituent minerals, some of which may be further altered. The feldspar in granite, for example, can be converted into clay by weathering processes. The grains resulting from weathering and erosion can be transported by running water, ice or wind and deposited as sediments in the sea, a lake or elsewhere on the land surface. In time these become sedimentary, secondary, rocks. Through depth of burial and the movement of the Earth's crust resulting from plate tectonic events, rocks are changed by heat and pressure to become metamorphic rocks. These in turn, if they are buried deep enough, may melt and create new magma to start the cycle again.

We are reliant on rocks, and the minerals they contain, for much of what we need. They provide our raw materials, from metals to fuel and chemicals, and our land is fertilised for crop production with chemicals derived from rocks. However, the Earth's crust has been unsustainably plundered and the rock cycle is being changed as the climate is altered by excessive use of fossil fuels.

IGNEOUS ROCKS

Igneous rocks form when magma or lava crystallises. Magma is liquid rock below ground (intrusive rock); lava is liquid rock on the surface (extrusive rock). Both contain the chemical components from which a range of silicate minerals develops during cooling. The temperature of the magma or lava and the situation in which it consolidates determine the characteristics of an igneous rock.

CLASSIFYING IGNEOUS ROCKS

Grain size and texture

Magma may solidify deep underground as a large mass. Such bodies (igneous intrusions) are called plutons or batholiths and can be many kilometres across. As cooling is very slow because of the depth of the intrusion and its great size (it can take tens of millions of years for a batholith to solidify), large crystals develop. Minor intrusions, such as sills and dykes, cool much more rapidly and the crystals in their rocks are smaller. A sill is a sheet-like

Basalt lava flows with ropy surface structures, Thingvellir, Iceland.

Contact between pinkish granite (right) and dark hornfels, Rinsey, Cornwall, England.

structure which occurs parallel to (concordant with) the existing strata, whereas a dyke is a more or less vertical sheet which cuts across (is discordant with) existing structures. When lava is erupted, it cools very rapidly, so extrusive rocks have small crystals which can often only be seen with a hand lens or microscope. Igneous rocks are subdivided and classified according to their crystal (grain) size. Coarse-grained rocks have crystals larger than 5mm in diameter, fine-grained rocks have crystals less than 0.5mm in diameter, and medium-grained is between the two, at 0.5 to 5mm in diameter. The grains in a coarse-grained rock are readily seen with the naked eye, but in medium- and fine-grained rocks a hand lens (x10 magnification) is needed.

Grain size is one of the characteristics of rock texture. The term texture refers to the size, shape and orientation of the constituent grains in a rock and the relationships between them. Sometimes certain crystals develop much larger than those in the rest of the rock that surrounds them. This can happen when the temperature at which the larger crystals form is maintained for a long period.

For example, magma below the surface can cool slowly and so large crystals form. This magma may then be erupted as lava which will cool rapidly into small crystals. The resulting rock will have a fine-grained matrix enclosing larger crystals. This common texture is referred to as a porphyritic texture and the large crystals as phenocrysts. Another common texture occurs when all the grains are the same size; this is an equigranular texture and is formed by uniform cooling.

Volcanic rocks may have textural features resulting from gas in the original lava. As lava cools and gas escapes, hollows and bubble-shaped cavities may remain in the solidified lava, representing places where gas bubbles existed. Such cavities are called vesicles, and solidified lava containing many of these is referred to as vesicular lava. Vesicles are ideal sites for the growth of minerals. An amygdaloidal texture is one with minerals infilling the vesicles.

Pillow lavas formed by underwater eruption, Anglesey, North Wales.

Columnar jointing in basalt, caused by shrinkage during cooling of lava, Fingal's Cave, Staffa, Scotland.

Composition

The other main characteristic of an igneous rock is its composition. This is determined by the type of original magma and the way in which the magma or lava has cooled. The terms 'acid' and 'basic' are taken from the language of chemistry and used in igneous petrology to refer to the amount of silica in the rock. The vast majority of the minerals of which igneous rocks are composed are silicates and quartz (silicon dioxide). Common igneous rock-forming minerals are feldspar, mica, olivine, pyroxene and amphibole. The proportion of these determines whether the rock is rich in silica (acid rock) or poor in silica (basic rock). Between these two categories are the intermediate rocks, and with a lower silica content than the basic rocks is the ultrabasic group. Magma generated very deep in the crust or in the upper mantle, especially in oceanic regions where the Earth's crust is relatively thin, tends to have a basic composition, whereas magma in the very thick continental crust is usually acidic. Acid rocks with more than 65% total silica are rich in mica and feldspar and have more than 10% quartz (often as much as 30%). An identification point is that acidic igneous rocks are pale in colour. They have a low specific gravity of around 2.6. Basic rocks have between 45% and 55% total silica and less than 10% quartz. They also contain plagioclase feldspar and pyroxenes such as augite.

Sheet intrusion of pink acid pegmatite in dark ultrabasic rock, Oldshoremore, Scotland.

Olivine can also be common. Basic rocks are dark in colour and have a specific gravity of about 3.0. Intermediate rocks can have features of both these groups. The ultrabasic rocks are rich in ferromagnesian minerals such as pyroxene, amphibole and olivine, which give them a specific gravity approaching 3.5.

Grain size	Composition			
	Acid Over 65% silica >10% quartz	Intermediate 55%–65% silica	Basic 45%–55% silica <10% quartz	Ultrabasic <45% silica
Coarse >5mm	Granite Pegmatite	Syenite Diorite Larvikite	Gabbro	Peridotite Serpentinite Dunite
Medium 5mm to 0.5mm	Microgranite	Microsyenite	Dolerite	
Fine <0.5mm	Rhyolite Obsidian	Andesite	Basalt	

GRANITE

Composition Granite is an acid igneous rock. It contains more than 10% quartz, often as much as 30%. The total silica content is over 65%. Common minerals include both biotite and muscovite mica, feldspar (more commonly orthoclase than plagioclase) and hornblende. Some granites contain accessory minerals, including pyrite, beryl, tourmaline and fluorite. This mineral content gives granite a pale overall colouring, and if there is much pinkish or red-coloured orthoclase feldspar, this will give the rock an overall reddish colour.

Grain size/texture Granite is a coarse-grained rock, with crystals over 5mm in diameter. It is often an equigranular rock, but many granites have much larger crystals set into the rock matrix, giving it a porphyritic texture. The large crystals (phenocrysts) are commonly of feldspar.

Occurrence This common igneous rock forms in large-scale igneous intrusions, originally a few kilometres under the surface. These plutons and batholiths generally develop in the roots of mountain chains and granite is only exposed on the surface after much weathering and erosion has taken place.

Granite from Cumbria, England, showing large pink feldspar crystals set in a matrix of smaller grey quartz, black biotite mica and pale feldspar crystals, giving the rock an attractive porphyritic texture.

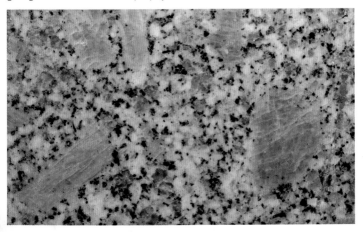

Pale granite from Cornwall, England, showing porphyritic texture with white orthoclase phenocrysts, granular grey quartz and black biotite mica.

Uses Granite is much quarried for use in the construction industry. Many types of granite, especially those with a porphyritic texture, are used decoratively for the facades of buildings and for domestic worktops; the rock takes an attractive polish. The word 'granite' is, however, much misused, to mean just about any stone kitchen work surface or other decorative rock.

Pinkish granite from South Africa. The pink crystals are orthoclase feldspar, with grey vitreous quartz and black biotite mica.

MICROGRANITE

Composition Microgranite has an acid composition, containing over 65% total silica and more than 10% quartz. There is more orthoclase feldspar than plagioclase, and other minerals include mica and hornblende. The overall colour of the rock is pale, but it may be pinkish if red- or pink-coloured orthoclase is present.

Grain size/texture Microgranite is a medium-grained igneous rock, with crystals ranging from 0.5mm to 5mm in diameter. Although generally having an equigranular texture, with randomly interlocking crystals, microgranite may be porphyritic. When the porphyritic texture contains phenocrysts of quartz, the rock is called quartz porphyry.

Occurrence As with other medium-grained rocks, microgranite forms in minor igneous intrusions, especially sills and dykes, and around the margins of larger intrusions.

Uses Microgranite may be used decoratively and is quarried in some areas as building stone and aggregate.

Medium-grained microgranite from France; pinkish orthoclase feldspar and some dark biotite mica visible.

Pegmatite composed of very large pink orthoclase crystals with pale grey quartz. Specimen from Sutherland, Scotland.

PEGMATITE

Composition Pegmatite is an acid igneous rock with a granitic composition. There is over 10% quartz, together with high proportions of orthoclase feldspar and mica. Hornblende also occurs. Pegmatite may be named according to a dominant mineral, for example feldspar pegmatite. Large crystals of various minerals, including tourmaline, topaz, beryl, zircon and apatite, can be found in pegmatite. These form from the fluids associated with the magma from which the rock crystallises.

Grain size/texture This rock has very large crystals, far bigger than those of the other coarse-grained igneous rocks. Often the crystals in pegmatite are many centimetres long.

Occurrence Pegmatite occurs in veins and dykes associated with large bodies of magma, where fluids from the later stages of magmatic consolidation are concentrated. The slow cooling in such situations allows time for very large crystals to grow.

Uses Pegmatite can be a source of lithium-bearing minerals, including spodumene. Other rare elements may be present, such as caesium and tantalum. Beryllium, which can be alloyed with nickel and copper, is obtained from beryl in pegmatites. Gem-quality topaz, fluorite, corundum, tourmaline and beryl are found in this rock.

DIORITE

Composition Diorite is a rock of intermediate composition, having between 55% and 65% total silica with some quartz and much plagioclase feldspar. Other minerals include biotite mica, amphibole (often hornblende) and pyroxene (commonly augite). Such mineralogy gives diorite an overall greyish colour.

Grain size/texture This is a coarse-grained igneous rock, the crystals being more than 5mm in diameter. It is usually equigranular, though some diorites have phenocrysts of feldspar or other minerals and are porphyritic.

Occurrence Diorite forms around the margins of large intrusions and often in small bodies, including veins and dykes. It can also occur as part of larger intrusions of both acid and basic composition. This rock is intruded in volcanic and mountain regions.

Uses Because of its durability, diorite is used for paving and roadstone. It takes an attractive polish and has been used decoratively.

Medium-grained, grey-coloured diorite with some crystals of dark pyroxene and amphibole. Specimen from Cumbria, England.

Greyish coarse-grained syenite with white feldspar and dark ferromagnesian minerals. Specimen from Scotland.

SYENITE

Composition Syenite has an intermediate composition with between 55% and 65% total silica. Quartz is present, but only up to around 10%. Mineral content consists of both orthoclase and plagioclase feldspar, with orthoclase in a much higher percentage, along with mica, amphibole and pyroxene. Other minerals include diopside and sodalite. Syenite is medium- to pale-coloured and can look rather like granite. However, a close inspection of the mineral content and the presence of a more varied mineralogy with pyroxene indicates that this is an intermediate rock and not acidic granite.

Grain size/texture This is a coarse-grained rock, usually of equigranular texture. The component mineral crystals are readily seen with the naked eye, though a hand lens will be needed to distinguish them accurately. Syenite is less frequently porphyritic than granite.

Occurrence Syenite can occur associated with intrusions of granitic rocks, but tends to form in smaller bodies than granite, for example in intrusions that have cooled slowly at some depth.

Uses As an attractive crystalline rock, syenite is widely used as a decorative stone for the facades of buildings, kitchen worktops and similar purposes. It is often incorrectly referred to as 'granite' when used decoratively. Syenite is also quarried for the same uses as many other igneous rocks, including roadstone and construction work.

LARVIKITE

Composition Larvikite has an intermediate composition, with 55% to 65% total silica. It has a similar mineralogy to syenite, containing pyroxene, amphibole and feldspar. Larvikite is characterised by its content of blue-grey plagioclase, which often shows a distinctive lustrous play of colour called schillerization. Olivine is an accessory mineral, with occasional apatite.

A polished specimen of larvikite from Norway. The bluish feldspar crystals are well displayed.

Grain size/texture This is a coarse-grained igneous rock, with crystals greater than 5mm in diameter. The plagioclase crystals are often over a centimetre long and occur in patches in the rock matrix.

Occurrence Larvikite usually occurs in relatively small, slow-cooling intrusions that form at some depth.

Uses An attractive rock, especially when polished, larvikite is much employed decoratively for the fronts of buildings. The incorrect trade name 'blue granite' is often used.

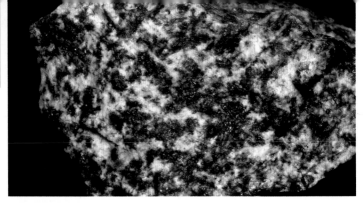

The two main minerals in coarse-grained basic gabbro are pale plagioclase and dark pyroxene. Specimen from Isle of Skye, Scotland.

GABBRO

Composition Gabbro is a basic rock, with less than 55% total silica and very little quartz (less than 10%). There is a more simple mineralogy than that of the acid and intermediate rocks, often with two dominant minerals, pyroxene (usually augite) and plagioclase feldspar, making up the bulk of the rock, with small amounts of quartz and possibly olivine. If much olivine is present, the rock may be called olivine gabbro. Magnetite (iron oxide) is a common accessory mineral. This mineralogy gives the rock a darker overall colour than the acid and intermediate rocks, with a noticeable dark and light speckled appearance.

Grain size/texture This is a coarse-grained rock, with crystals over 5mm in diameter. The great majority of gabbros are equigranular. Layering in gabbro is not uncommon, with alternate layers of dark and lighter minerals. This is usually due to settling of denser and lighter minerals influenced by gravity during the cooling of magma.

Occurrence As with many of the rocks of basic composition, gabbro is associated with oceanic crust and magma generated in the Earth's mantle. Its coarse grain size indicates slow cooling in large masses such as thick dyke sheets and plutonic bodies.

Uses Gabbro is quarried as an ornamental rock for paving and for facing buildings. 'Black granite' is a common trade name for gabbro used for tombstones and kitchen work surfaces. A variety of accessory minerals of economic value found in some gabbros includes nickel, cobalt, platinum and silver.

DOLERITE

Composition A rock of basic composition, dolerite has the same mineralogy as gabbro and basalt. It contains two main minerals, plagioclase feldspar and pyroxene (often augite), with less than 10% quartz. Olivine may be present, along with magnetite (iron oxide). The overall silica content is less than 55%, and the rock is dark-coloured, grey or greenish-black, often with a speckled appearance. Diabase is the North American term for dolerite.

Grain size/texture Dolerite is a medium-grained rock, with crystals between 0.5 and 5mm in diameter. Generally it has an equigranular texture, but phenocrysts may occur, giving a porphyritic texture.

Occurrence Dolerite is a classic rock of sills and dykes. Dyke swarms, where a great number of related dykes occur, are common around many volcanic centres, such as the Tertiary dykes of the Inner Hebrides, Scotland, and the Karoo dykes of South Africa. Dyke swarms such as these expand the Earth's crust by considerable amounts. The Whin Sill, which intrudes Carboniferous strata in northern England, and the Palisades Sill in New Jersey, USA, are dolerite concordant sheet intrusions.

Uses Dolerite is a very durable rock and as such is extensively quarried for roadstone, railway ballast and other constructional purposes.

A dark basic igneous rock, dolerite contains pale plagioclase and dark pyroxene. Specimen from Durham, England.

A dark-coloured ultrabasic rock, this specimen of peridotite has a groundmass of dark ferromagnesian minerals with small reddish garnet crystals.

PERIDOTITE

Composition Peridotite is an ultrabasic igneous rock, containing less than 45% total silica. This rock is composed mainly of dark, dense ferromagnesian minerals such as olivine, pyroxene or amphibole. Garnet is another common constituent. Pale, lower-density minerals such as feldspar are absent. Olivinite (dunite), pyroxenite and amphibolite are varieties of peridotite named according to their dominant mineral.

Grain size/texture This is a coarse-grained rock, with the constituent crystals being more than 5mm in diameter. Most peridotites are equigranular and some have a layered structure.

Occurrence The composition of peridotite is possibly very similar to that of the Earth's upper mantle. Some peridotites probably formed as a result of the cooling of magma generated in the mantle. Other peridotites result from the accumulation of dense minerals produced early in the consolidation of basic magma. They are often associated with the oceanic crust.

Uses Many peridotites are rich in olivine and gem-quality olivine crystals (peridot) occur. A report in 2008 suggested that peridotite may have considerable use in capturing carbon dioxide from the atmosphere, thus reducing the effects of pollution in causing global warming. Peridotite is known to react with carbon dioxide to form calcareous rocks, thus storing this greenhouse gas.

DUNITE

Composition Dunite is composed almost entirely of olivine, but there may be small amounts of garnet and pyroxene. The total silica content is less than 45% and quartz is absent. Because of this mineralogy, it has also been called olivinite. It belongs to the peridotite group of ultrabasic igneous rocks.

Grain size/texture This is a medium-grained rock, with crystals between 0.5 and 5mm in diameter. Its texture is equigranular and it can have a rather 'sugary' appearance.

This dunite from New Zealand is composed almost completely of greenish and greyish olivine.

Occurrence Dunite has a composition thought to be very similar to parts of the upper mantle of the Earth. It is probably derived from basaltic magma. Olivine, the main constituent of dunite, is one of the very first minerals to crystallise as magma cools. Dunite can therefore form by the settling of olivine crystals through still-molten magma, to create a layer of olivine-rich ultrabasic rock. This has happened in the Palisade Sill of New Jersey, USA.

Uses Dunite has no important economic uses, but the rock may be associated with deposits of chromium.

A pale-coloured, flinty specimen of rhyolite from North Wales. The rock is too fine-grained for the minerals to be detected.

RHYOLITE

Composition Rhyolite is a pale-coloured acid igneous rock containing over 65% total silica and more than 10% quartz. It has the same composition as granite and quartz porphyry. The essential minerals in rhyolite are quartz, feldspar (the orthoclase percentage being greater than that of plagioclase), mica and hornblende. Glass (very rapidly cooled lava with a high silica content and no real crystal form) is a major constituent.

Grain size/texture This rock forms from rapidly cooled lava. Such rapid cooling leads to the development of few crystals, much of the rock having a fine-grained or glassy texture. Even with microscope examination, the constituent minerals may be difficult to determine. The flow of such viscous lava may give rise to a banded structure, picked out by lighter and slightly darker bands.

Occurrence Rhyolite is the result of very explosive volcanic eruptions and the cooling of viscous lava. Such lava may even consolidate in a volcanic pipe and violent explosions are needed to produce further eruptions.

Uses Rhyolite is used in road making and other construction work, but its very high silica content prevents it from being of use as a concrete aggregate.

OBSIDIAN

Composition Obsidian is an acid igneous rock with more than 65% total silica and a very high percentage of quartz. It has the same mineralogy as granite, but much of its composition is glass, lacking well-formed crystals because of very rapid cooling. Obsidian may contain about 1% water. The rock is usually black in colour, but 'snowflake obsidian' has numerous white flecks, which are patches of devitrified glass and cristobalite (a relative of quartz).

Grain size/texture This is a very fine-grained rock, with no well-formed crystals. The term anhedral is used to refer to poorly formed crystals. With a glassy texture, obsidian breaks to give very sharp edges and a curved, conchoidal fracture. Even with microscopic examination, only small, roughly formed minerals can be seen.

Occurrence Obsidian forms where acid lava has cooled very rapidly and is associated with rhyolite flows.

Uses Some types of obsidian, including the snowflake variety, are used ornamentally. In the distant past its sharp edges were exploited as cutting tools.

Black, glassy obsidian from California, USA, showing vitreous lustre and sharp fractured edges.

PITCHSTONE

Composition This rock has a variable composition, but there is commonly a high silica content and a mineralogy similar to that of granite and rhyolite. Some varieties have a composition not unlike that of gabbro and basalt. Many pitchstones are hydrated types of obsidian.

Grey pitchstone from the Isle of Arran, Scotland, with typical greasy lustre and jagged fracture.

Grain size/texture Pitchstone is a very fine-grained rock, rather like obsidian, though it contains more identifiable crystals. However, it may include some phenocrysts. It is not as dark-coloured as obsidian and, as its name suggests, it has a pitch-like lustre. Pitchstone varies from grey to greenish, red or black in overall colour. Many pitchstone exposures exhibit flow banding.

Occurrence This rock forms by the very rapid cooling of lava, usually in flows, but it also occurs in some minor intrusions.

Uses There are no modern widespread uses of pitchstone, though in the past, together with obsidian, it has been used for making tools and weapons because of its hardness and sharp-edged fracture.

ANDESITE

Composition This is an igneous rock of intermediate composition, having between 55% and 65% total silica. Its mineralogy is similar to that of syenite, with much plagioclase feldspar, along with dark mica, amphibole and pyroxene. The rock may be grey, brown or greenish in colour. It is a volcanic equivalent of diorite and syenite.

Vesicular andesite from the volcano Poas in Costa Rica. Medium-coloured, fine-grained groundmass with phenocrysts of pyroxene.

Grain size/texture Andesite is a fine-grained volcanic rock, the individual grains being less than 0.5mm in diameter. The rock is frequently porphyritic, with small phenocrysts of feldspar or ferromagnesian minerals. Andesite may have an amygdaloidal or vesicular texture.

Occurrence A very widespread volcanic rock, andesite is associated with subduction zones and island arcs. Its name comes from the Andes Mountains. It occurs in lava flows and is present in violently erupting volcanoes. The rock has also recently been found in meteorites and in the Martian crust.

Uses Andesite is used locally for construction and road making.

Dark-coloured, fine-grained vesicular basalt with a large olivine phenocryst. Specimen from Iceland.

BASALT

Composition Basalt is a very common basic igneous rock with 45% to 55% total silica content and a low percentage of quartz (less than 10%). Its dark colour, often almost black, is a result of a high proportion of ferromagnesian minerals, especially pyroxene (usually augite). Nearly half the rock is made up of plagioclase feldspar, and olivine is often present. Magnetite is an accessory mineral. Basalt is the volcanic equivalent of gabbro and dolerite.

Grain size/texture This is a fine-grained lava, with crystals only visible through a hand lens or microscope. Some basalts have phenocrysts of pyroxene or olivine, giving the rock a porphyritic texture. Vesicular and amygdaloidal textures are common, and the rock is then called vesicular or amygdaloidal basalt. Amygdales can be of zeolite minerals and quartz. Large vesicles may contain agate.

Occurrence Basalt occurs in oceanic areas where it flows freely from volcanic vents. Because it is a very fluid lava, basalt flows great distances and forms extensive sheets. The Earth's ocean floors are composed of basalt and submarine eruptions may form pillow lavas (see p. 12). These result when a skin of solidified lava develops on contact with cold sea water and molten lava is erupted into the skin to produce a rounded 'pillow' shape. Basalt is a major constituent of the Moon's surface and is also present on Mars.

Uses Basalt can be used in the construction industry. Because of its magnetite content and magnetism relative to the changes in the Earth's magnetic poles, studies of sea floor basalt have been important in working out the way in which the ocean crust is moving.

AGGLOMERATE

Composition Agglomerate is a volcanic rock composed of materials that have come from the vent of a volcano. These can include blocks of lava and other pyroclastic material, together with fragments of country rock caught up in the explosive activity. Agglomerate may resemble the sedimentary rock conglomerate, but a study of the fragments it contains will show these to be of volcanic origin.

Grain size/texture This rock consists of coarse and fine particles deposited together and usually unsorted. The larger particles tend to be held in a finer ashy matrix. Agglomerate formed near to a volcanic vent may be welded by lava. Volcanic bombs and twisted spindle-shaped fragments can also occur.

Occurrence Agglomerate is produced in volcanic craters or very close to them. Some agglomerates may show features rather like sedimentary stratification.

Uses There are no known significant uses.

RIGHT: Coarse angular fragments of various rock types set in a reddish matrix. Specimen from Italy. BELOW: Volcanic 'slag' from Iceland with a red iron-rich content. Agglomerate is a consolidated version of this material.

ROPY LAVA

Composition Because it flows very readily, basic lava often forms ropy structures. It contains much plagioclase feldspar and pyroxene, with some olivine and magnetite, and is dark brownish or black in colour.

Dark basaltic lava with a ropy surface structure indicating flow when molten. Specimen from Hawaii.

Grain size/texture Ropy lava is a fine-grained rock, which is very fluid when erupted because of a high gas content. This gives rise to a notably vesicular texture, with numerous small gas bubble cavities. The ropy surface structures are the result of flow of the molten lava. Often this happens after a crust or skin has developed and the still-molten lava flows beneath this solidified upper layer, pushing it into ridges.

Occurrence Ropy lava occurs in many parts of the world, especially in the vast shield volcanoes of the ocean regions. In Hawaii this type of lava is called pahoehoe and this local word has become part of the scientific language of geology.

Uses There are no known significant uses.

Fine-grained tuff from North Wales showing various small inclusions.

TUFF

Composition Tuff is a pyroclastic rock composed of material blown from a volcanic crater. Some tuffs are composed of small fragments of rock, while others may be mainly made of glassy or crystalline material. If mineral components can be determined, they are those typical of explosive volcanic activity and may include pyroxene, plagioclase feldspar and amphibole. Some tuffs, formed by pyroclastic flows, are welded by small droplets of lava which were suspended in the flow. These flows are among the most devastating of volcanic eruptions. The tuffs they produce (ignimbrites) are usually of acid composition.

Grain size/texture This is a fine-grained rock, with ash and some slightly larger fragments set into the finer matrix. If tuff is deposited into water, it may have bedding characteristic of a sedimentary rock.

Occurrence When volcanic ash and dust are blown from a volcano, this material can be taken high into the atmosphere and may even be carried far round the Earth, creating vivid sunsets. The thickest deposits of tuff are found near to the parent volcano. Volcanic dust which has been carried a great distance and deposited is very useful as a stratigraphic marker.

Uses Tuff has been widely used as a building stone. In the USA, a storage repository for spent nuclear fuel is situated in a deposit of tuff at Yucca Mountain, Nevada.

METAMORPHIC ROCKS

There are a number of ways in which previously formed rocks can be metamorphosed (changed). Their original igneous, sedimentary or existing metamorphic features can be totally altered to produce a new rock type. These changes take place within the Earth's crust as a result of heat, pressure, a combination of both, or migrating fluids. At temperatures lower than 200°C, little change takes place, metamorphism mainly occurring at temperatures above this and up to around 700°C. Pressures increase with depth in the Earth's crust and at around 20km are 6,000 times atmospheric pressure. There can be some overlap between the processes involved, but metamorphic rocks are generally divided into three groups for the purposes of understanding and identification; these groups reflect how the metamorphic rock was formed.

CONTACT METAMORPHISM

When magma or lava (at as much as 1,000°C) comes into contact with pre-existing rocks in, or on, the Earth's crust, this heat alters the cold crustal rocks. Contact or thermal metamorphism brings about recrystallisation, so that original structures are removed. New minerals may grow within the altered rock and, from a study of these, geologists can work out temperatures and sequences of metamorphism. Heating and change are greatest immediately next to the magma or lava, and metamorphism gradually decreases away from the igneous rock. Different areas of metamorphism are recognised according to the minerals in the changed rock. A giant batholith will have a wide metamorphic aureole (zone) of possibly a few kilometres around it. The effect of a small dyke may only be measured in centimetres. The type of igneous rock causing the metamorphism has a great bearing on how far the aureole extends and the type of changes. High-temperature fluids (often water-based) are associated with granitic magma and these fluids can invade the country rock, bringing about chemical changes. Some rocks resist metamorphism more than others; clay and shale are readily altered, whereas igneous rocks and quartzites may change very little.

Cliff of gneiss, with alternating acid and ultrabasic bands, Sutherland, Scotland.

REGIONAL METAMORPHISM

This type of metamorphic change occurs during periods of mountain formation, usually associated with plate tectonics. Rocks caught up in folding and often buried deep in the Earth's crust are altered by both heat and pressure during such events. Regional metamorphism may affect rocks over extensive areas, and a number of grades of regional metamorphism are identified, related to increasing pressure and temperature conditions.

Low grade

At this grade pressure is low and temperature is hardly an influence. Very few rocks are altered except those least resistant to change, such as clay and shale. Slate is a typical metamorphic rock of low-grade regional metamorphism, its main characteristic being slaty cleavage. This is a structural feature brought about by minerals in the rock being aligned by pressure conditions. Low-grade rocks tend to be fine-grained.

Medium grade

Both pressure and temperature are influences here and many rocks are changed. The typical medium-grade metamorphic rock is schist,

which is characterised by a wavy foliation or banding. The rock has a sheen because of the growth of much mica during metamorphism, and new minerals including garnet and kyanite form.

High grade

Virtually all preformed rocks are changed by this grade of regional metamorphism, which involves the very highest heat and pressure conditions, without rock being melted. The high-grade metamorphic rock gneiss is characterised by a coarse grain size and alternating bands of light and dark minerals. Fluids circulating through the crust are also responsible for some of the metamorphic changes, by introducing or removing mineral components.

DYNAMIC (DISLOCATION) METAMORPHISM

This is a less common way in which rocks are changed. Dynamic metamorphism occurs near the fault plane of a large-scale thrust fault, when vast masses of rock are moved. The rocks near the fault plane are altered by crushing. Pressures and temperatures are usually low and the majority of rock changes are due to brecciation (breaking). Different mineral grains in the original rock are altered in different ways. Those with low resistance to crushing are ground into very small fragments, while tougher material may survive to become porphyroblasts set in the finer matrix. Mylonite, with its streaked-out appearance, is the typical rock of this type of metamorphism.

ABOVE: Dark hornfels, Cornwall, England.

RIGHT: Disused marble quarry, Iona, Scotland.

BELOW: Regional metamorphism occurs deep beneath mountain ranges, as here at The Remarkables in Otago, South Island, New Zealand.

White calcite-rich marble with yellowish-green colouring due to the presence of brucite formed during metamorphism. Specimen from Sutherland, Scotland.

MARBLE

Composition Marble (metamorphosed limestone) is a highly calcareous rock, composed mainly of calcite. During metamorphism impurities in the original limestone are changed into new minerals by heat and migrating fluids. Because of its high calcite content, marble is a pale-coloured rock, varying from white to grey. There is a wide range of minerals which develop during metamorphism and these can give marble distinctive colours. Diopside causes greenish colouring and brucite adds green or bluish veins and patches. Olivine, wollastonite, tremolite and serpentine minerals may also occur.

Grain size/texture This contact metamorphic rock has a crystalline texture, with individual grains varying from fine to coarse. When viewed with a lens, marble is seen as an interlocking mosaic of calcite crystals. The original limestone would have had pore spaces and fossil material, but both these features are usually removed by metamorphism.

Occurrence Marble forms near to igneous intrusions and lava flows, when direct heat from the magma or lava changes the original limestone. The metamorphic changes are most profound nearest to the igneous rock and the metamorphism gradually fades with distance away from the heat source.

Uses This attractive rock has been used decoratively and for construction for thousands of years. Many marbles are named after their source locality, for example, Vermont marble from the USA and Carrara marble from Tuscany, Italy. Marble is widely used for flooring and worktops.

HORNFELS

Composition Hornfels can have a varied mineralogy depending on the original rock and the type of magma providing the heat. This is a quartz-rich rock, which also contains much biotite mica, feldspar and pyroxene. Cordierite, andalusite and garnet are also commonly formed during heating. Some hornfelses have thin bladed crystals of chiastolite (a variety of andalusite) and there may be spots visible on the rock surface, also composed of andalusite.

Dark hornfels with porphyroblasts of pale andalusite. Specimen from Brazil.

Grain size/texture Generally a fine- to medium-grained rock, hornfels is crystalline and has a flinty appearance. A granoblastic texture may be present, in which crystals of minerals such as garnet and andalusite are set into the finer-grained matrix of the rock.

Occurrence Hornfels occurs very close to igneous intrusions, especially those of large size.

Uses Hornfels is locally quarried for road aggregate and similar uses.

METAQUARTZITE

Composition Metaquartzite has a very high percentage of quartz, often more than 90%. Other minerals include feldspar and mica. The rock is usually white or pale grey in colour, though the presence of iron oxide minerals may give it a darker or pinkish colour. Metaquartzite forms when sandstone is altered by contact with magma or lava.

Grain size/texture This is a medium- to fine-grained rock with a mosaic of fused quartz crystals. The original quartz grains in the sandstone would have had pore spaces between them, but metamorphism causes these to be infilled as the quartz crystals grow. Metaquartzite is a 'sugary' rock, which often lacks structures related to the original sedimentary rock.

Occurrence Metaquartzite forms near to igneous intrusions where heating is sufficient to affect the resistant quartz grains in sandstone. Some metaquartzites occur during high grades of regional metamorphism, when temperatures and pressures are great.

Uses Because it is composed of quartz crystals, metaquartzite is a hard rock. It is therefore widely quarried for road making and railway ballast and is used as a source of industrial sand.

Sugary metaquartzite composed almost entirely of pinkish-grey quartz. Some hints of original sedimentary bedding remain.

Greyish slate, with large porphyroblasts of pyrite. Specimen from Cumbria, England.

SLATE

Composition This is a rock formed by low-grade regional metamorphism and so only a few rocks such as shale and clay are altered. Slate is therefore mainly composed of quartz and mica, with feldspar, chlorite group minerals, clay minerals and graphite. Pyrite is a common accessory mineral forming discrete crystals. Slate is coloured by included minerals. Green slate contains chlorite; dark grey colouring comes from graphite or disseminated pyrite. Fossils are sometimes found in slate, indicating how little the composition of the rock has been altered. These include graptolites, brachiopods and trilobites, although they may be distorted by metamorphism.

Grain size/texture Slate is a fine-grained rock, with most of the grains the same size. Porphyroblasts (larger crystals set in the matrix of the rock) of pyrite are common. The main feature of slate is its cleavage (slaty cleavage) brought about by minerals in the rock becoming aligned by the pressure of metamorphism. This is different from the original sedimentary bedding of the rock and slaty cleavage often cuts through any relict bedding.

Occurrence This rock is found in areas where mountain formation has occurred; it forms mainly in the outer regions of such mountain belts.

Uses Slate's main economic use is related to its property of cleavage. Because slate splits readily along its cleavage planes, often into thin layers, it has been used for hundreds of years for roofing. The rock is also used for flooring, tiling and other constructional purposes.

SCHIST

Composition Schist generally contains much mica, together with quartz and feldspar. Minerals in the rock produced by the pressure and temperature of metamorphism include garnet, kyanite, hornblende, pyroxene, glaucophane and chlorite. Particular schists are named after their mineral content, for example garnet schist. Schist is often a silvery rock, with mica giving the rock a distinctive sheen on its foliation surfaces.

Grain size/texture This is a medium-grained rock, composed of well-formed crystals. An important feature of the rock is the wavy banding structure (schistosity) running through the rock. The minerals, especially mica, are aligned along this structure, which may cause the rock to split. Minerals in schist can occur in bands, some rich in mica, others in quartz and feldspar. This segregation of minerals is not as distinct as in the high-grade rock, gneiss.

Occurrence Schist forms in mountain belts, at deeper levels than slate. Here pressure and temperature are moderate and many original rock types are changed. A variety of sedimentary rocks, including shale, clay, sandstone and limestone, together with some igneous rocks and the metamorphic rock slate, can be altered to form schist.

Uses There are no known significant uses.

Silvery schist coloured by large amounts of mica. Wavy foliation gives the specimen an undulating surface.

GNEISS

Composition Gneiss (pronounced 'nice') can have a composition not unlike that of the igneous rock granite. It contains quartz, feldspar and mica, with hornblende. Garnet and pyroxene may also be present.

Grain size/texture A coarse-grained rock with a banded texture, gneiss contains crystals which are easy to see with the naked eye. The rock is characterised by alternating dark and light bands made of different minerals. The pale bands contain quartz and feldspar, while the darker bands are predominantly biotite mica and hornblende.

Occurrence This rock forms deep in the roots of mountain belts where temperatures and pressures are extreme and any preformed rock is altered. Eventually in these situations melting may occur and magma is then formed. Areas containing gneiss are often of extreme age. The Canadian Shield and the Lewisian rocks of north-west Scotland contain gneisses that may represent parts of the original continental crust of the Earth. Some of these rocks were metamorphosed around 3,000 million years ago.

Uses Locally used as building stone and aggregate. Because of its resistance to weathering and erosion, gneiss is used for coastal defence work.

Grey banded gneiss from Sutherland, Scotland. The darker bands are rich in ferromagnesian minerals, the pale bands in quartz and feldspar.

ECLOGITE

Composition Eclogite is composed of ferromagnesian minerals, mainly garnet and pyroxene. It therefore has a low silica content (and high density) and is similar in composition to an ultrabasic igneous rock. Because of its composition, eclogite is a striking rock, with green and red mottling. Kyanite, rutile, glaucophane and diamond can occur in eclogite.

A coarse-grained eclogite from Norway. The green mineral is olivine, the reddish mineral is garnet.

Grain size/texture This is a coarse-grained rock, with crystals obvious to the naked eye. Eclogite can have a banded texture, but is often a crystalline rock without noticeable structures.

Occurrence Eclogite forms at the very highest grades of metamorphism and is close to being an igneous rock. It occurs at the deepest levels of the crust where conditions are extreme, and can be found with peridotite (an ultrabasic igneous rock) and serpentinite. It may result from the metamorphism of basic igneous rocks such as basalt and gabbro.

Uses There are no known significant uses.

SERPENTINITE

Composition As its name indicates, this rock is made up of minerals belonging to the serpentine group. These include fibrous antigorite and chrysotile. It also contains many ferromagnesian minerals such as pyroxene, garnet and hornblende. This mineralogy gives the rock a dark greenish colour, often mottled with grey, blue and red. It has the silica content of an ultrabasic igneous rock.

Serpentinite from Cornwall, England. This large specimen shows the typical banded colours of the rock.

Grain size/texture Serpentinite is a crystalline, medium- to coarse-grained rock, which may have vein-like bands of colour.

Occurrence This rock is found where basic and ultrabasic igneous rocks have been altered by the processes of serpentinisation. This metamorphic change involves water as well as heating and possibly occurs deep in the crust below the ocean floor. Some authorities include serpentinite in the igneous rocks, but essentially it is a rock that results from the metamorphism of a preformed rock.

Uses Serpentinite, because of its attractive colouring and veining, is carved into various decorative objects. It also has properties which arrest radiation and so is used in the nuclear industry in concrete and radiation shielding.

MYLONITE

Composition Mylonite has a very variable mineralogy depending on the rocks involved in its formation. Much of the rock consists of pulverised rock dust, together with recrystallised minerals, some of which may have formed after alteration.

Fine-grained mylonite from Ross and Cromarty, Scotland, showing alternating layers of dark and pale rock material.

Grain size/texture This is a fine- to medium-grained rock, which is recognised by its 'streaked-out' appearance. Porphyroblasts of minerals such as garnet occur in some mylonite. There may be lenses of rock material within the rock's matrix.

Occurrence Mylonite forms near the fault plane of a large-scale thrust fault. Here considerable shearing stress occurs, grinding existing rocks to a rock 'flour' and drawing the particles out in the direction of fault movement. Temperatures are low. Thrusting of this type occurs during mountain formation, when two opposing rock masses come together and one rides over the other.

Uses Locally used as aggregate.

SEDIMENTARY ROCKS

Sedimentary rocks are readily distinguished from igneous and metamorphic rocks by their stratification (bedding). This is easily recognised in a rock exposure as neat layering. In a small hand specimen this may be less easy to see, but a sedimentary rock will break along its bedding planes to determine the shape of the fragment. Stratification is a result of how a sedimentary rock forms and the processes involved. The majority of these rocks are created by weathering and erosion of pre-formed rocks, transportation of the grains by a river system, ice or the wind, and finally deposition in water or on the land surface. The particles deposited in water form neat strata. Because sedimentary rocks form on or near the Earth's surface, the processes involved in their creation are readily observable, unlike those which form the majority of igneous and metamorphic rocks. Another important feature of sedimentary rocks is that they contain fossils which are our evidence of organic evolution. Very few metamorphic rocks have fossils in them and crystalline igneous rocks have none.

Stratification in shales and limestones, Kimmeridge Bay, Dorset, England.

The various chemical and physical changes which take place in sediments in order to make them into rocks are referred to as diagenesis. Porosity generally decreases and new minerals, which cement grains together, may be introduced.

These rocks are identified according to their texture (the shape, size and relationship of their grains) and their composition. Three main groups of sedimentary rocks are recognised: detrital, organic and chemical rocks.

DETRITAL SEDIMENTARY ROCKS

This large category includes the well-known sandstones, shales, clays and conglomerates. These are made of fragments (detritus) broken from previously formed rocks by weathering and erosion, two processes often misunderstood and confused, but very different.

Weathering

This is the breakdown of rock materials without any movement. Mechanical weathering involves processes including the build-up of ice in rock joints and pore spaces and the subsequent stress set up in the rock. Heating and cooling of rocks, especially in arid regions, may cause exfoliation (peeling of 'onion skin' layers on rock surfaces). Chemical weathering involves acid waters. Rain is

Mechanical weathering producing scree, Wastwater, Cumbria, England.

Chemical weathering has reduced the level of this limestone pavement since the glacial erratic was deposited around 10,000 years ago, leaving the erratic on a small plinth. A glacial erratic is a piece of rock which had been carried some distance by an ice sheet or glacier and deposited when the ice melts. North Pennines, England.

naturally weak carbonic acid, having taken up carbon dioxide from the atmosphere. Many other acids, available through man-made pollution, increase the rain's acidity. Calcareous rocks are those most affected by acid water, though feldspar and other minerals are also broken down by the effects of acidic waters. Acid water combines with insoluble calcium carbonate in limestone and converts it to calcium bicarbonate, which is dissolved and removed. Feldspar is changed to clay minerals, which are readily washed away. Biological weathering by lichens, fungi and animals can lead to the initial breakdown of rocks.

Erosion

Erosion is the breakdown of rocks involving movement. This occurs in rivers, glaciers, and by the action of the sea and the wind. Many complex processes are involved, including abrasion, hydraulic action and plucking. Rock fragments are broken down by contact with each other and the bed of a river or sea cliff. Gravity is also important in cliff development and can cause mass movement down slopes. Erosion takes place during transport to a site of deposition, where sediments settle to become sedimentary rock.

ORGANIC SEDIMENTARY ROCKS

This group includes rocks which have a high content of organic material or have been formed by organic processes. Many limestones are organic, as they contain much shell or other fossil material. Coral-rich limestones may represent original reef environments cemented by lime mud. Coal, formed mainly of carbon derived from plants, is an organic rock of high economic value.

CHEMICAL SEDIMENTARY ROCKS

Inorganic chemical activity produces some sedimentary rocks, including oolitic limestones, where calcium carbonate is precipitated around small nuclei in warm agitated sea water. Rock gypsum and rock salt are formed through chemical precipitation in evaporite basins when bodies of saline water dry up, and certain ironstones are formed chemically.

SEDIMENTARY STRUCTURES

Sedimentary rocks, especially those classified in the detrital group, frequently exhibit structures related to their formation. These are very useful when geologists are trying to work out the environment in which the rock formed, as structures can often be compared with modern features.

Bedding planes may display ripple marks and desiccation cracks. Ripples indicate moving water, such as the ripples formed on the seashore or a riverbank. Desiccation cracks (mud cracks) have roughly hexagonal outlines on a bedding surface and show that this area was above water level when the sediment dried and cracked. A group of structures referred to as sole or groove marks are the result of objects being bounced or dragged along a soft sediment surface. These are often caused by deep ocean currents.

Some sedimentary structures are internal rather than on bedding planes. Cross bedding, where bedding planes may appear curved and cut across each other in section, results from deposition in moving water. A large-scale type of cross bedding, dune bedding, is wind deposited and resembles very closely the structures seen in modern sand dunes. Certain sediments have size grading of the grains within a bed. This graded bedding usually has coarser grains at the base and gradually finer grains towards the top of the bed. Turbidity (density) currents, which flow over the ocean floor, commonly deposit graded beds. The coarse sediment is carried in the base of the current and the finer material, held in suspension, settles later. Each current deposits one such bed and many can be found in a rock succession. These beds are called turbidites after the turbidity currents that deposit them.

ABOVE: Ripple marks preserved on sandstone bedding plane, Stoer Bay, Sutherland, Scotland.

RIGHT: Coastal erosion with cliffs and arch, Durdle Door, Dorset, England.

BELOW: The Twelve Apostles limestone stacks, Victoria, Australia, created by marine erosion.

Red conglomerate from Devon, England, containing rounded pebbles of various materials held in a sandy matrix. The red colouring is due to haematite.

CONGLOMERATE

Composition Conglomerate is made of rounded rock fragments cemented together. The type of fragments depends on the type of rock in the source area. Often there is a predominance of quartz fragments in the rock, but it may contain quite a variety of different rock materials.

Grain size/texture This is a very coarse-grained rock, often containing pebble- to boulder-sized fragments. These are frequently cemented by finer material and possibly quartz or calcite. The larger grains are rounded. Bedding structures are often not visible in hand specimens, though they may be apparent in the field.

Occurrence Conglomerate forms in a number of different environments. These are characterised by having enough energy to move large pebbles and other coarse-grained sediment. Such environments include beaches and fast-flowing river systems. Conglomerates are often the lowermost deposit following a period of erosion and rest unconformably on older eroded rocks. An unconformity may represent an old land surface, and as sea level rises, conglomerate is formed, possibly as a beach deposit. The roundness of the included grains is a strong indication of movement and erosion by water and often rapid deposition, as the fragments have not been reduced much in size.

Uses When weathered or fragmented, conglomerate may be an excellent source of gravel and aggregate.

BRECCIA

Composition Breccia is made of large fragments cemented by a finer-grained matrix. The composition of the fragments, as with conglomerate, depends on their source area. Breccias may be named after the majority of the fragments, for example, quartz breccia or limestone breccia.

Large and small angular fragments of quartz-rich rock set in a reddish matrix.

Grain size/texture The large fragments in breccia are angular, as is the cementing material. There may be a considerable difference in size between the large and smaller fragments, and the rock is thus said to be poorly sorted. The fragments are generally randomly orientated and bedding may be absent.

Occurrence Breccia forms in a number of environments, but is rarely associated with running water. It commonly occurs in regions in which mechanical weathering is dominant. The rock is likely to be formed by the accumulation of scree at the base of a cliff. Some breccias are associated with faulting and volcanic activity.

Uses Breccia is sometimes used as an aggregate.

SANDSTONE

Composition Sandstone contains a high proportion of quartz in the form of small- to medium-sized grains. These may be cemented by quartz or calcite, and some sandstones have reddish iron oxide coating their grains. Other minerals include glauconite, which, when present in a high percentage, can give the rock an overall greenish colour (greensand). Feldspar is common in some sandstones, the rock then being called arkose. Mica occurs in many sandstones and appears as small glittery flakes on the bedding planes. If the rock contains a very high quartz content, it is referred to as orthoquartzite. This is a very mature sedimentary rock, as opposed to those which contain a greater variety of minerals (especially feldspar), which are immature. This concept indicates the amount of processing the grains have undergone from source area to deposition, as quartz is one of the most resistant minerals and therefore survives considerable erosion.

Grain size/texture Sandstone is a medium-grained rock, with all grains much the same size and a well-sorted texture. If the grains

Red sandstone coloured by haematite and composed of rounded wind-blown grains of quartz.

are angular, it is likely that they have been influenced by water action, but if rounded ('millet-seed grains'), they have probably been moved and deposited by the wind in an arid environment. Bedding structures, including cross bedding caused by deposition in moving water or wind, are readily visible in field exposures.

Occurrence Sandstone occurs in a great variety of environments. Millet-seed sandstone is a classic arid desert deposit, while quartz-rich rocks with angular grains come from river or marine environments. Sandstone containing mica is unlikely to have formed in an arid environment, as small flakes of mica are easily removed by the wind. Arkose, an immature feldspar-rich sandstone, probably forms near to its source area by rapid deposition, as feldspar is easily broken down by weathering.

Uses These rocks are much quarried for the construction industry, both as building stone and sand. Some sandstones are an important source of water, as their porosity allows them to be underground aquifers. Oil and gas are frequently trapped in sandstone.

Sandstone slabs on the summit of Cul Mor, north-west Scotland.

SHALE

Composition Shale contains much quartz, with clay minerals, mica and feldspar. Gypsum and pyrite crystals are not uncommon; dark shales are coloured by high proportions of organic material. Shale commonly contains well-preserved fossils, though some of these may be crushed on the bedding planes.

Grain size/texture This is a fine-grained, well-sorted rock, with all the grains the same size. Shale is neatly bedded and readily splits into thin layers. The rock can contain rounded nodular masses, which have an overall composition similar to that of the rock itself. These can have concentrations of calcite or pyrite and may contain perfectly preserved fossils.

Occurrence Shale commonly forms in water which lacks much movement and energy, such as deep oceanic areas well away from sources of coarse detrital sediment. The fossil content will be a reasonable indication of the depositional environment. Some shales, however, form in shallower conditions.

Uses Certain shales, known as kerogen shales, contain sufficient amounts of oil to make them economically useful. The shale, once mined or quarried, can be heated to produce oil and gas. Recently the hydraulic fracturing (fracking or fraccing) of shales containing oil and gas has been used as a method of winning valuable reserves.

Grey shale composed of fine-grained detrital sediment along with organic material. Specimen from North Yorkshire, England.

Grey boulder clay containing small rock fragments.

CLAY

Composition Clay is made up of detrital quartz, with feldspar, mica, chlorite group minerals and clay minerals. These clay minerals are silicates, containing elements such as aluminium, magnesium and iron, with water. Typical clay minerals are illite, kaolinite and montmorillonite. They are derived from the weathering of feldspar and other silicate minerals and commonly occur as fine-grained particles. Clay can contain well-preserved fossils.

Grain size/texture This is a very fine-grained rock, with grains all much the same size. The individual grains cannot be seen, even with a hand lens. When dry, clay cracks and breaks easily, with a powdery feel, but it becomes plastic when wet.

Occurrence Clay forms in a variety of ways. It can occur in deep marine conditions and in lakes. The fossil content will be useful to determine the exact conditions. Clay is also a product of glacial activity. Boulder clay is produced by the grinding action of ice sheets and the rock debris they contain and is a typical glacial deposit. Certain clays, called varved clays, are formed in lake deposits associated with glacial conditions. They consist of alternating dark and paler layers. The pale, thicker layers represent the warm season, with more sediment from meltwater, the thinner, dark layers the cold seasons.

Uses This rock has a variety of uses and is quarried for the construction industry, especially for brick making and the manufacture of cement. It is also the basis of the ceramics industry.

The soft surface of this marl specimen from southern England shows numerous fractures.

MARL

Composition Marl is similar to clay in many respects, containing much detrital quartz, clay minerals and silt. However, marl is defined by its high calcite content, often between 40% and 60%. Two colour varieties of marl commonly occur. Red marl is coloured by iron oxide; green marl contains minerals such as chlorite and glauconite (a complex hydrous silicate mineral).

Grain size/texture Marl is a very fine-grained rock, with all grains the same size. Calcite can act as a cement, holding the grains together. The rock often shows bedding structures, or, like clay, it may be very finely laminated.

Occurrence Marl occurs in both marine and freshwater environments. Green marls containing glauconite are marine in origin, whereas red marls are often associated with evaporite deposits and dolomite (dolostone), and are found with halite, potash salts and gypsum.

Uses There are no known significant uses.

LIMESTONE

Composition Limestone contains a very high percentage of calcium carbonate (calcite), which gives the rock a pale colour. There may also be a certain amount of detrital quartz and other sediment. The calcite in many limestones is of organic origin and these rocks are classified accordingly. Shelly material derived from molluscs or brachiopods can be dominant; some limestones are full of coral fragments while others contain broken crinoid stems. The individual rock is named with reference to the fossil content, for example, coral limestone or crinoidal limestone. Calcite can also be precipitated chemically from seawater. This occurs during constant agitation in warm water when layers of calcite are formed around small nuclei to produce ooliths.

Grain size/texture The grain size varies with different types of limestone, but is generally medium- to coarse-grained. Oolitic limestones are well-sorted rocks made of rounded ooliths about 2mm in diameter. The fragments of fossil material in limestone may be a few centimetres in length. The matrix of many limestones is fine-grained lime mudstone.

Occurrence The majority of limestones are marine rocks, formed in relatively shallow water. Their fossil content indicates the conditions of deposition.

Uses Limestone is used in the construction industry, as a raw material for cement and concrete, and as a flux in steel making.

LEFT: *An organic limestone with fossils of crinoids, brachiopods, bryozoans and corals, cemented by hardened lime mud. Specimen from Shropshire, England.*
RIGHT: *Pale oolitic limestone composed of small, rounded, calcareous ooliths held in a calcite matrix. Specimen from Gloucestershire, England.*

CHALK

Composition Almost entirely composed of calcite, chalk is made of the remains of micro-organisms, especially coccoliths and foraminiferans. A very small amount of detrital material, mainly clay and silt, is present. Chalk has a white, powdery appearance and is porous. It is an organic sedimentary rock.

Grain size/texture Chalk is fine-grained, microscope examination being required to determine its detailed composition. It is a bedded rock, the stratification often picked out by layers of black flint nodules. These are composed of cryptocrystalline silica and probably formed on the seabed before the chalky ooze became rock.

Occurrence This extremely pure calcite rock originated as marine lime ooze derived from the accumulation of the remains of micro-organisms. It has been suggested that the lack of detrital material in chalk is due to any nearby land areas being low-lying and arid.

Uses Chalk is quarried for use in cement making and traditional lime mortar. In some areas it is used as a building stone. Because it is porous, chalk can hold water. Beneath London the synclinal (downfolded) chalk strata act as an aquifer.

White, powdery, very fine-grained chalk. Specimen from Hampshire, England.

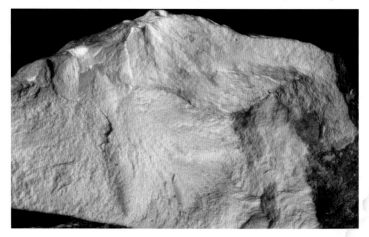

DOLOMITE

Composition Dolomite is also called dolostone (an American term), to prevent confusion with the mineral dolomite. It is often referred to as magnesian limestone. There is a considerable percentage of the mineral dolomite in this rock, together with calcite and detrital material, often quartz and clay minerals. Dolomite is generally darker than calcite-rich limestones, and often has a brownish colour. It is classified as a chemical sedimentary rock.

Grain size/texture This is a crystalline sedimentary rock, with dolomite crystals cemented by lime mud. Bedding structures may be lacking; the term 'massive' is used to describe sedimentary rocks which are unbedded. Dolomite may contain fossils, but they are far less common than in calcite-rich limestones. In the field, exposures may show reef structures and large, nodular masses.

Occurrence Dolomite rock is generally thought to have originated as limestone, with recrystallisation and alteration of calcite to dolomite. This can take place when percolating fluids seep through limestone in a marine environment and in arid regions of high salinity.

Uses This rock is quarried extensively as a source of magnesium and as a flux in the steel industry. It is also used as a fertiliser and in the construction industry, for the manufacture of heat-resistant bricks and as aggregate.

Brownish-cream dolomite from County Durham, England. The specimen shows some chemical weathering.

IRONSTONE

Composition There are a number of different ironstones in the Earth's crust, but the majority contain haematite, magnetite or goethite. Among the most important of these, both from a geological and economic perspective, are the banded ironstones found in many areas. These rocks are very ancient, having been formed over 3,000 million years ago. It has been argued that their deposition was due to the atmosphere becoming enriched with oxygen from bacterial photosynthesis and iron being oxidised. The darker layers are rich in iron oxides and the reddish layers are silica-rich chert. Some haematite-rich ironstones are formed by the replacement by haematite of original minerals in the rock, especially calcite in some limestones. Oolitic ironstones are also of economic importance. They may be formed when chemical changes alter the rock after diagenesis and are thus classified as chemical sedimentary rocks.

Grain size/texture Ironstones are generally medium- to fine-grained rocks, with a well-sorted texture. Oolitic rocks have a mass of small rounded grains.

Occurrence Oolitic and limestone-based ironstones may be formed in marine environments, whereas banded ironstones occur in continental areas. Some magnetite deposits accumulate in beach and river sands.

Uses Ironstones have been the basis of the iron and steel industry for many years. Some are excavated in vast open-cast pits, while many, because of their high value, are mined deep underground. Important centres of ironstone extraction are in China, Australia, India, Russia, Brazil, the USA, Canada, South Africa and Sweden. There are good deposits in the UK, but these are no longer exploited because of the high cost of production compared with elsewhere.

ABOVE: Banded ironstone from Western Australia. Alternating dark, iron-rich layers and pale quartz layers. BELOW: Haematite-rich ironstone from Sierra Leone.

Bituminous coal with typical dull, dusty areas and brighter reflective patches. Specimen from South Africa.

COAL

Composition Coal is an organic sedimentary rock made of carbon derived from plant material. There may be a very small amount of detrital sediment in the rock and the purity of the coal is determined by the total carbon content. Anthracite is a high-carbon coal which burns at high temperatures; bituminous coal is of lower carbon content and has a higher volatile percentage.

Grain size/texture This rock is fine- or medium-grained. Plant fragments are visible in some types of coal.

Occurrence Coal forms from the alteration of peat deposited by the accumulation of vegetable matter. Much of the economically useful coal is derived from thick forest peat deposits. Over time these were buried under a great thickness of other sediment and subsequently heated. This heating drove off volatile material, especially water, increasing the carbon content. Coal usually occurs as discrete beds (seams) in an alternating sequence of strata with shale and sandstone.

Uses This rock is one of the most valuable and has been used for hundreds of years as a source of heat. It is a primary fuel for the generation of electricity; nearly 70% of China's power is from this source and the USA approaches 50%. Coal-fired power stations are inefficient and produce vast amounts of atmospheric pollution, causing climate change and acid rain. The mining of coal is hazardous and the waste from coal-burning power stations contains uranium and arsenic in its fly ash.

FLINT

Composition Flint is composed of cryptocrystalline silica, possibly derived from organic sources such as sponge spicules. It forms as discrete rounded nodules and is a type of chert. Flint is very hard and breaks with a curved, conchoidal fracture to leave extremely sharp edges. Fossils, especially of echinoids and molluscs, are frequently found in flint nodules.

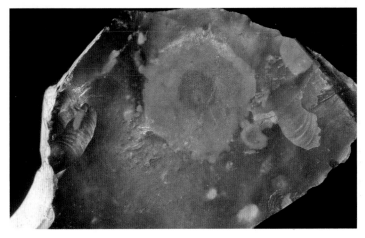

Dark broken flint nodule with white chalky coating and conchoidal fracture. Specimen from Norfolk, England.

Grain size/texture This is a crystalline rock with no individual grains. It may have a white chalky coating but the internal colour is generally dark grey or black. Some nodules contain cavities filled with earthy material; others may have chalcedony fillings.

Occurrence Flint is found as layers in chalk. These show up as dark bands in contrast to the white colouring of the chalk. Flint's exact origin is not fully understood. It may be derived from mobile silica which became trapped in animal burrows and other cavities in the chalky sediment before hardening to flint.

Uses Flint, because of its hardness and sharp fracture, was used for thousands of years for tools and weapons. Its main use today is as a local building stone.

Iron oxides colour bauxite red. This unbedded specimen from Surinam shows characteristic fine-grained texture.

BAUXITE

Composition Bauxite is often classified with minerals, but as it is an aggregate of several minerals and has a varied chemistry, it can be considered a rock. It mainly consists of hydrated oxides of aluminium and iron oxides. These minerals include diaspore, gibbsite, boehmite and limonite, the last giving a brownish-yellow colour. Haematite and goethite are also usually present, together with clay minerals.

Grain size/texture This rock is usually unbedded (massive). Oolitic and concretionary bauxite deposits are known. The grain size varies from fine to medium.

Occurrence Bauxite is formed by the alteration, especially in tropical conditions, of preformed rocks containing silicates of aluminium. These are then leached away, leaving bauxite-forming minerals.

Uses Bauxite is the main ore of aluminium and is mined especially in Russia, Indonesia, Jamaica and Surinam. Aluminium is a light, strong metal resistant to alteration in the atmosphere. It is extensively used in many ways, from aircraft to drinks cans and household devices. It has an increasing use in the motor industry, as car bodies are made lighter to save fuel.

Crystalline rock salt composed mainly of halite. Specimen from North Yorkshire, England.

ROCK SALT

Composition Rock salt is mainly made of crystalline halite (sodium chloride), fused in a mosaic of interlocking crystals. There is also a variable amount of detrital sediment in the rock, including silt, marl and small quartz fragments. Rock salt is variable in colour depending on the impurities. The rock is often pinkish or red in colour because of the presence of iron oxide (haematite). Much silt will colour the rock grey.

Grain size/texture This is a crystalline, chemically formed sedimentary rock, often without stratification (massive). The mineral halite forms perfect cube-shaped crystals and these may be visible in specimens of rock salt. The rock has a greasy feel and will dissolve in water. It is very soft and can be easily marked with a fingernail.

Occurrence Rock salt forms in evaporite basins, which may be dried-up inland salt lakes or marine areas, such as lagoons, isolated from the main body of the ocean. As the salt-rich waters dry out, rock salt is formed, interbedded with other evaporites, marl and detrital sediments.

Uses Rock salt has a great many uses. It is important in the chemical and food industries and vast amounts are put on the roads in northern latitudes to prevent ice forming in winter.

ROCK GYPSUM

Composition This rock is a stratified form of gypsum (hydrated calcium sulphate). The colour varies according to impurities. Much rock gypsum is white or grey, but reddish varieties are coloured with haematite (iron oxide).

Grain size/texture Rock gypsum is a crystalline, chemically formed rock. The surface often shows a mass of gypsum crystals fused together, with a silky sheen. It is easily scratched with a fingernail.

Occurrence This is an evaporite sediment, produced with other evaporites in dried-out lake and sea areas. It forms interbedded with rock salt and potash rock, and with marl, limestone and sandstone. It is readily folded and often occurs in wavy layers.

Uses Rock gypsum has many uses, especially in the chemical industry and in the manufacture of plaster and plasterboard.

A fragment of rock gypsum coloured by iron oxide. Specimen from Cheshire, England.

MINERALS

Minerals are the material components of which rocks are made. They have structural and chemical properties which can be defined, and with very few exceptions minerals are solid and inorganic. Some, such as gold and diamond, are composed of atoms of single elements, but the vast majority are chemical compounds.

Each mineral is defined and identified by certain properties, most of which are easy to observe. When trying to identify a mineral, each property should be considered. After this has been done, an 'identikit' list for the mineral will have been made and by comparing this with known properties, the identification of the mineral can be worked out. Some of the properties used for identification are observations (colour and habit, for example), while others, such as hardness and specific gravity, can be measured using reasonably simple tests.

MINERAL SHAPE

Minerals often form beautiful crystals, but they commonly occur in irregular shapes. There are a number of crystal systems (shape categories) into which crystals can be placed. These crystal systems are defined according to symmetry. The habit of a mineral is the actual shape of the specimen and defining this habit is a very valuable identification property. A number of everyday and scientific words are used to describe different crystalline and non-crystalline habits.

Prismatic	With a constant cross-section
Tabular	Tablet-shaped
Bladed	Like knife blades
Acicular	Needle-shaped, often in a radiating mass
Dendritic	With a tree-like shape
Reniform	Kidney-shaped
Botryoidal	Like a bunch of grapes
Mammilated	As botryoidal, but larger scale
Massive	With no definable shape; an irregular specimen

Words such as 'cubic' and 'hexagonal' can be used to describe crystal habits.

Sometimes two or more crystals grow together and share a common crystal surface. Such crystals are said to be twinned and may radiate from a central point, penetrate each other or lie back to back.

Rather than going into the detailed symmetry of each crystal system, it is more useful here to mention typical shapes (habits) that each system has.

Crystal systems

Cubic	Cubes, octahedra (8 sides) and dodecahedra (12 sides)
Tetragonal	Prisms, more elongated crystals than in the cubic system
Orthorhombic	Tabular and prismatic shapes
Monoclinic	Prisms and tabular shapes
Triclinic	Prisms and tabular shapes
Trigonal/hexagonal	Prisms, tabular and rhombohedral shapes

The trigonal and hexagonal systems are so similar that they are usually grouped together.

THE WAY MINERALS BREAK – CLEAVAGE AND FRACTURE

Not all mineral specimens are perfect unbroken examples. The vast majority of specimens show broken surfaces and the way a mineral breaks is a reasonably good identification property. There are two ways in which a mineral may break: it may cleave, to produce even surfaces, or it may fracture in an irregular way. Most minerals fracture, but not all minerals cleave.

Cleavage

This is a product of the internal atomic structure of the mineral. Surfaces along which the mineral cleaves are often related to layers of atoms and so the cleavage of a certain mineral can be repeated many times over. Though not as smooth as crystal surfaces, cleavage planes can reflect light in a consistent way. Words used to describe a mineral's cleavage include perfect, distinct and indistinct. To observe that the mineral has no cleavage can be a useful identification clue.

Minerals often exhibit fine crystal habit and rich colours like these prismatic crystals of microcline feldspar, variety amazonite.

Fracture

Unlike cleavage, this property is not closely related to the internal atomic structure and therefore when a mineral fractures it has irregularly broken surfaces. 'Uneven' is a common word used to describe fracture. 'Hackly' describes an uneven fracture with sharp jagged edges and a 'conchoidal' fracture is curved. The fracture of a mineral cannot be repeated exactly, as cleavage can.

COLOUR

The colour of a mineral in natural light can be a valuable identification feature and is one of a mineral's most immediately obvious characteristics. However, there are certain problems when using colour as an identification property. Many individual minerals exhibit a wide variety of colours. The common oxide mineral quartz can occur in a great many colour forms. These are often so striking that they have individual names: amethyst is purple and rose quartz is pink, for example. Also, many different minerals are the same colour, often white or other pale colours.

STREAK

The streak of a mineral is its colour when powdered. Usually the powder is consistent for any given type, even though that mineral may exhibit a variety of colours. So the streak of purple quartz, pink quartz, white quartz and all the other coloured quartzes is always white. Streak is obtained by drawing the mineral across an unglazed tile (the streak plate). With very hard minerals it may be necessary to use a harder surface, or crush a small fragment of the mineral with a geological hammer.

LUSTRE

This is the way light is reflected by the mineral's surface. There are a number of terms used to describe different lustres. These include: adamantine (brilliant, sparkling), vitreous (glassy), dull, metallic, silky, pearly, splendent (glittery), resinous and greasy.

HARDNESS

This is a property that requires a simple test to be carried out on the mineral under investigation. Hardness is the resistance to scratching on the mineral surface. A standard scale, the Mohs' hardness scale, is used and a mineral's hardness is measured against this scale. Not all the points on the scale are an equal measurement apart, but it has common minerals as its points and so has proved to be useful.

In order to test a mineral's hardness, a scientific approach should be adopted, beginning by trying to scratch the specimen with talc and continuing up the scale until the mineral no longer resists being scratched, when its hardness will have been found. If, for example, the mineral is not marked by calcite but is scratched by apatite, its hardness will be 4.

Mohs' Hardness Scale

```
1  Talc
2  Gypsum
3  Calcite
4  Fluorite
5  Apatite
6  Orthoclase feldspar
7  Quartz
8  Topaz
9  Corundum
10 Diamond
```

Mineral vein cutting through red sandstone. The vein contains pale quartz and red haematite. Fife, Scotland.

There are also certain everyday objects that can be used for hardness testing. A fingernail is 2½ and a steel knife blade 5½. It should be possible to obtain most of the minerals on this scale. Hardness testing kits are available. Hardness is always written as a number and a fraction; decimal points are not used, to avoid confusion with specific gravity.

SPECIFIC GRAVITY (SG)

Scientifically this is a comparison between the weight of the mineral and that of an equal volume of water. This test is not always possible to carry out, as often a mineral may be attached to others or to rock. With experience it is apparent whether a mineral is heavier than average. The average SG of many minerals is around 2.0 to 3.0. Gold, with an SG of 19.3, is far higher than the rest. Quartz, a common mineral, has an SG of 2.65. It should be relatively easy to say which of two equal-sized mineral specimens has the higher SG. (Note the use of decimals with this property.)

SPECIAL PROPERTIES

Certain minerals may exhibit properties such as magnetism, fluorescence under ultraviolet light, phosphorescence, double refraction, radioactivity and reaction with liquids.

MINERAL FORMATION

Many minerals occur in veins cutting through the Earth's crust. Mineral veins often follow fault lines or joint planes, and very fine mineral

Economic conditions determine the viability of mining. This is the abandoned tin mine at Wheal Coates, Cornwall, England, which was worked in the 1800s and early 1900s.

crystals can be found in these. Hot fluids, rising under pressure from depth, seep into these fractures and minerals are formed from them. These hydrothermal fluids may contain a wide variety of mineral-forming elements and they are often associated with granite intrusions. Sulphides such as galena and sphalerite occur in hydrothermal veins, as do fluorite, barite, quartz, calcite and gold.

Igneous rocks are formed either from lava on the surface or magma which crystallises below ground. As these rocks cool and solidify, minerals such as feldspar, mica, quartz, hornblende, augite and olivine form. Where cavities occur in lava or magma, huge crystals can develop.

Metamorphic rocks contain a range of minerals which develop during metamorphism of existing rocks. These minerals include garnet, kyanite, mica, pyrite, hornblende and feldspar.

Sedimentary rocks are often made of fragments of mineral broken off previously formed rocks. Some, including limestone and ironstone, contain minerals which are formed at the same time as, or soon after, deposition of the rock. Minerals such as calcite, limonite, haematite, quartz and dolomite can form between the grains of a sedimentary rock, thus helping to cement the rock particles together. Evaporite minerals, including halite and gypsum, occur in sedimentary deposits.

ABOVE: *Cut and polished sapphires on rough sapphire.*

RIGHT: *Some minerals have special properties. This rhomb of calcite (Iceland spar) shows double refraction.*

BELOW: *Two prismatic emerald crystals on pale quartz, Bolivia.*

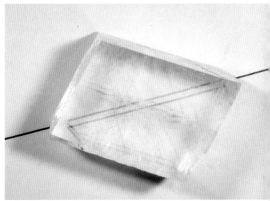

NATIVE ELEMENTS

This group of minerals consists of uncombined elements, both metals and non-metals.

GOLD

Group	Native Elements
Composition	Au
Crystal System	Cubic
Habit	Crystals are very rare, cubes and octahedra; grains, nuggets, flakes, dendritic shapes
Cleavage	None
Fracture	Hackly
Specific Gravity	19.3
Hardness	2½ to 3
Colour	Rich yellow
Streak	Golden-yellow
Lustre	Metallic

Gold occurs in hydrothermal mineral veins, often with quartz. Because of its very high specific gravity, it becomes concentrated in alluvial sands as a placer deposit. There are more than 10 million tonnes of gold in the oceans and though various methods of obtaining this supply have been suggested, it has yet to be exploited commercially. The main mining regions are in the USA, Canada, South Africa, Russia and Peru. Gold is insoluble in acids, except in aqua regia and selenic acid. Water heated to 375°C under high pressure will also dissolve it. This valuable metal has been used ornamentally for thousands of years, as it is malleable and easily shaped. It also serves as a standard for international currency.

Small nuggets and grains of gold on hexagonal milky quartz prisms. Specimen from South Africa.

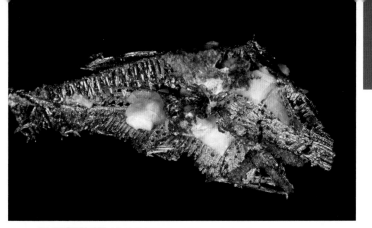

Short wires of silver on calcite. Specimen from Australia.

SILVER

Group	Native Elements
Composition	Ag
Crystal System	Cubic
Habit	Crystals are very rare, cubes and octahedra; massive, as scales, wires and dendritic shapes
Cleavage	None
Fracture	Hackly
Specific Gravity	10.5
Hardness	2½ to 3
Colour	Silvery-white
Streak	Silvery
Lustre	Metallic

Silver is found with gold, and with minerals containing silver, in hydrothermal veins. When ore deposits are altered and oxidised, silver may remain in the changed rock. When it occurs with gold, an alloy, electrum, is often formed. Silver is usually obtained from its ores, acanthite (silver sulphide) and stephanite (silver antimony sulphide) being important. The main producers are Mexico, the USA, Peru and Australia. Silver has been exploited for thousands of years for jewellery and other ornamentation. It is an excellent conductor of heat and electricity and has recently found a use in the disposal of chemical weapons. It is more soluble than gold and dissolves in nitric acid.

Nodular and massive copper with green malachite formed by alteration. Specimen from Peru.

COPPER

Group	Native Elements
Composition	Cu
Crystal System	Cubic
Habit	Crystals rare, cubes and octahedra; dendritic, massive, as wires
Cleavage	None
Fracture	Hackly
Specific Gravity	8.9
Hardness	2½ to 3
Colour	Coppery red, brown when tarnished
Streak	Reddish-copper
Lustre	Metallic

Copper often combines with other elements to form minerals such as chalcopyrite (copper iron sulphide). This mineral, rather than native copper, is the main industrial source of the metal and is predominantly mined in the USA, Australia, Canada, Chile and Indonesia. Copper forms as a native element in volcanic rocks of basic composition. This malleable metal has been used for thousands of years and is today essential for electrical wiring, piping and many other industrial applications.

SULPHUR

Group	Native Elements
Composition	S
Crystal System	Orthorhombic
Habit	Crystals tabular; massive, stalactitic, as powdery coatings
Cleavage	Imperfect
Fracture	Uneven or conchoidal
Specific Gravity	2.1
Hardness	1½ to 2½
Colour	Bright yellow
Streak	White
Lustre	Resinous

Sulphur is formed from fluids and gases issuing from hot springs and volcanic craters. It is mined in many areas, especially Texas and Louisiana, USA. However, one of the main sources of industrial sulphur is the hydrogen sulphide which has to be removed from natural gas. Sulphur has many industrial applications, such as the manufacture of sulphuric acid and in vulcanising rubber.

Crystals of sulphur with white calcite, from Italy.

DIAMOND

Group	Native Elements
Composition	C
Crystal System	Cubic
Habit	Cubes, octahedra, tetrahedra
Cleavage	Perfect
Fracture	Conchoidal
Specific Gravity	3.5
Hardness	10
Colour	Colourless, grey, white, yellow, pink, blue, green
Streak	White
Lustre	Adamantine or greasy

Diamond occurs in pipe-shaped igneous structures made of ultrabasic rock called kimberlite. Diamond defines point 10 on the hardness scale and is the hardest known mineral. The main regions where diamonds are mined are Russia, South Africa, Canada, Namibia, Botswana, Angola and Australia. Diamonds are used as gemstones and industrial diamonds are synthetically manufactured, mainly in Russia.

Small, 5mm, octahedral diamond crystal (just above and to right of centre) in kimberlite rock. Specimen from South Africa.

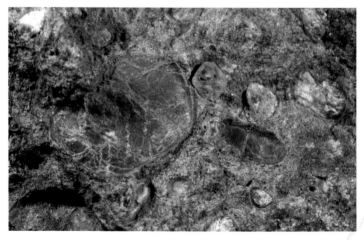

GRAPHITE

Group	Native Elements
Composition	C
Crystal System	Trigonal/hexagonal
Habit	Hexagonal, tabular; massive, granular and foliated
Cleavage	Perfect
Fracture	Uneven
Specific Gravity	2.2
Hardness	1 to 2
Colour	Black or very dark grey
Streak	Black
Lustre	Metallic

Graphite is found in schist and other metamorphic rocks such as slate. It is used in lubricants because of its chemical structure of sheets of carbon atoms, which gives specimens a greasy feel. In Cumbria, England, where graphite occurs in slate, it was the origin of a local pencil industry. Graphite is predominantly mined in China, Brazil, India, Canada and North Korea. Other uses include steel making, brake linings, refractories and batteries.

Massive graphite showing typical metallic lustre.

BISMUTH

Group	Native elements
Composition	Bi
Crystal System	Trigonal/hexagonal
Habit	Lamellar, massive, foliated, granular, dendritic
Cleavage	Perfect
Fracture	Uneven
Specific Gravity	9.8
Hardness	2 to 2½
Colour	Silvery, often tarnished red
Streak	Silvery
Lustre	Metallic

Bismuth occurs in hydrothermal mineral veins and in very coarse-grained pegmatites. It dissolves readily in nitric acid. Bismuth is used as an industrial catalyst. It is mainly produced not from the native element but as a by-product of copper and lead smelting. The chief producer is the USA and bismuth ores (chiefly bismuthinite) are mined in Bolivia, Canada, Japan, Mexico and Peru.

Silvery bismuth from the Czech Republic.

ANTIMONY

Group	Native elements
Composition	Sb
Crystal System	Trigonal/hexagonal
Habit	Tabular or pseudo-cubic, acicular; massive, lamellar, granular
Cleavage	Perfect
Fracture	Uneven
Specific Gravity	6.7
Hardness	3 to 3½
Colour	Grey, silver-grey
Streak	Grey
Lustre	Bright metallic

Antimony is found with sulphides such as galena, pyrite and sphalerite and with silver and arsenic in hydrothermal mineral veins. The main use of antimony is as an alloy with other metals for bearings, cable protection and batteries. It can also act as a fire retardant in plastics. China is the main source of antimony, with Russia, Bolivia and South Africa producing lesser amounts.

Massive, silvery-grey antimony, showing typical metallic lustre. Specimen from Mexico.

SULPHIDES

These minerals are chemical compounds formed by the combination of metals and semi-metals with sulphur. Many are important metal ores.

GALENA

Group	Sulphides
Composition	PbS
Crystal System	Cubic
Habit	Crystals cubic and octahedral; granular and massive
Cleavage	Perfect
Fracture	Subconchoidal
Specific Gravity	7.6
Hardness	2½
Colour	Grey
Streak	Grey
Lustre	Metallic

Galena is a common mineral of hydrothermal veins, where it is found with other sulphides and quartz, fluorite and calcite. It is an important ore of lead, with Australia, the USA and China producing much of the world's supply. Silver often occurs with lead (up to 1.2kg per tonne). The main uses of lead are in car batteries and as a radiation shield. When galena is dissolved in hydrochloric acid, hydrogen sulphide, smelling of bad eggs, is produced.

Mass of cubic galena crystals, some with twinning. Typical lead-grey colouring and metallic lustre. Specimen from Spain.

Granular cinnabar on rock matrix, showing characteristic colouring and vitreous lustre. Specimen from Spain.

CINNABAR

Group	Sulphides
Composition	HgS
Crystal System	Trigonal/hexagonal
Habit	Crystals rhombohedral, tabular, prismatic; granular, massive
Cleavage	Perfect
Fracture	Uneven or conchoidal
Specific Gravity	8.1
Hardness	2 to 2½
Colour	Brown to scarlet
Streak	Red
Lustre	Dull, metallic, vitreous, or adamantine

Cinnabar occurs near volcanic vents and hot springs, often with native mercury. Occasionally it is found in hydrothermal mineral veins. Cinnabar is the main ore of mercury and is produced in Russia, Spain, China and Italy. Because it is so toxic, many of its industrial uses are being taken over by other materials.

COBALTITE

Group	Sulphides
Composition	CoAsS
Crystal System	Orthorhombic
Habit	Crystals octahedral, with striated faces; massive, granular
Cleavage	Perfect
Fracture	Uneven
Specific Gravity	6.3
Hardness	5½
Colour	White to dark grey, blue
Streak	Dark grey
Lustre	Metallic

Cobaltite forms in some metamorphic rocks and in hydrothermal mineral veins in association with calcite, chalcopyrite, sphalerite, magnetite, skutterudite, titanite and zoisite. It is soluble in nitric acid. When melted, the globules formed are magnetic. Cobaltite, along with other cobalt ores, is mined for the metallic element. Erythrite can be produced when cobaltite is weathered.

Cobaltite crystals, with metallic lustre, on rock matrix. Some crystals show an octahedral habit. Specimen from South Africa.

Mass of sphalerite crystals on calcite. Typical material from a hydrothermal sulphide vein. Some crystals show a tetrahedral habit. Specimen from Durham, England.

SPHALERITE

Group	Sulphides
Composition	ZnS
Crystal System	Cubic
Habit	Crystals tetrahedral, dodecahedral; botryoidal, massive, granular
Cleavage	Perfect
Fracture	Uneven or conchoidal
Specific Gravity	3.9 to 4.1
Hardness	3½ to 4
Colour	Brown, yellowish, black, grey, reddish
Streak	Pale brown
Lustre	Resinous or vitreous

Sphalerite occurs in hydrothermal mineral veins, where it is commonly found with galena, quartz, pyrite, calcite and barite. Sphalerite is the main ore of zinc and is mined in the USA, Canada, Russia, Australia and Peru. An important industrial use is in galvanising steel and it is also used in batteries. Zinc oxide is important in the rubber industry.

STIBNITE

Group	Sulphides
Composition	Sb_2S_3
Crystal System	Orthorhombic
Habit	Crystals prismatic, with striated faces, bladed; granular
Cleavage	Perfect
Fracture	Uneven or subconchoidal
Specific Gravity	4.6
Hardness	2
Colour	Grey
Streak	Grey
Lustre	Metallic

Stibnite occurs around hot springs and in hydrothermal veins, with realgar, orpiment, galena, pyrite, barite, sphalerite, calcite and gold. It also forms in replacement deposits. The crystals may be twisted and bent. It will dissolve in hydrochloric acid. Stibnite is an important ore of antimony, mainly mined in China, Russia and Bolivia. Antimonite is an alternative name for this mineral.

Masses of slender, radiating, prismatic stibnite crystals with typical grey colouring and metallic lustre.

BORNITE

Group	Sulphides
Composition	Cu_5FeS_4
Crystal System	Cubic
Habit	Crystals rare, cubes, octahedra, dodecahedra; massive, granular
Cleavage	Poor
Fracture	Uneven or conchoidal
Specific Gravity	5.1
Hardness	3
Colour	Brown or coppery red, often tarnished purple, blue and red
Streak	Black
Lustre	Metallic

Bornite forms in hydrothermal mineral veins with galena and with other copper minerals including chalcopyrite. It very rarely occurs as crystals, generally being massive or as grains. Because of its brilliant colour when tarnished, bornite is often called 'peacock ore'.

Massive bornite of the form called 'peacock ore' because of its iridescent tarnish. Specimen from Cumbria, England.

CHALCOPYRITE

Group	Sulphides
Composition	$CuFeS_2$
Crystal System	Tetragonal
Habit	Crystals pseudo-tetrahedral, with striated faces; massive, reniform, botryoidal
Cleavage	Poor
Fracture	Uneven
Specific Gravity	4.4
Hardness	3½ to 4
Colour	Brassy yellow
Streak	Greenish-black
Lustre	Metallic

Chalcopyrite occurs in association with many other sulphides, including galena, pyrite and sphalerite, in hydrothermal mineral veins. When tarnished by exposure to the atmosphere, this mineral becomes iridescent. It is the main ore of copper and is extensively mined in Canada, the USA, Australia, Chile and Indonesia.

Massive, brassy-yellow chalcopyrite showing characteristic metallic lustre, with small amounts of milky quartz. Specimen from Cornwall, England.

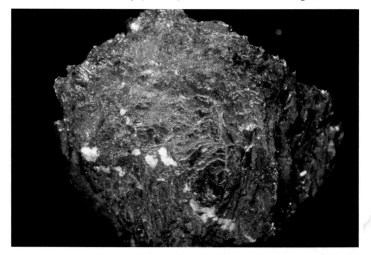

ORPIMENT

Group	Sulphides
Composition	As_2S_3
Crystal System	Monoclinic
Habit	Crystals rare, prismatic; foliated, massive
Cleavage	Perfect
Fracture	Uneven
Specific Gravity	3.5
Hardness	1½ to 2
Colour	Yellow to brownish-yellow
Streak	Pale yellow
Lustre	Resinous or pearly

Orpiment occurs at the margins of hot springs and in hydrothermal veins, often associated with realgar and stibnite. It also forms around volcanic vents and is a product of the alteration of realgar. When orpiment is heated, a smell of garlic is given off, because of its arsenic content. It dissolves in nitric acid. Widely used as a pigment and source of arsenic, other uses include the production of semi-conductors and infrared-transmitting glass.

Brownish-yellow orpiment crystals. Specimen from Peru.

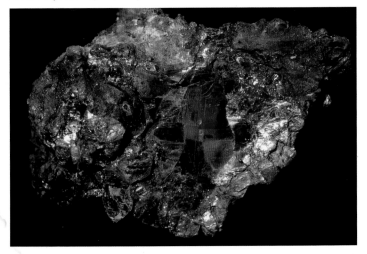

REALGAR

Group	Sulphides
Composition	AsS
Crystal System	Monoclinic
Habit	Crystals prismatic, with striated faces; massive, granular
Cleavage	Distinct
Fracture	Conchoidal
Specific Gravity	3.6
Hardness	1½ to 2
Colour	Red, orange-red
Streak	Yellow to orange-red
Lustre	Resinous

Realgar is found in hydrothermal mineral veins with orpiment and metallic sulphides. It also occurs near hot springs, where mineral-rich water crystallises. When heated, the characteristic garlic smell of arsenic is produced. If exposed to light for some time, realgar becomes an orange-yellow powder. This mineral is a source of arsenic.

Mass of small realgar crystals with characteristic orange-red colouring. Specimen from Namibia.

Scales of molybdenite showing characteristic dark grey colouring, metallic lustre and perfect cleavage.

MOLYBDENITE

Group	Sulphides
Composition	MoS_2
Crystal System	Trigonal/hexagonal
Habit	Crystals barrel-shaped or tabular; as scales, foliated
Cleavage	Perfect
Fracture	Uneven
Specific Gravity	4.6 to 5.1
Hardness	1 to 1½
Colour	Silvery grey
Streak	Grey
Lustre	Metallic

Molybdenite occurs in granite as an accessory mineral, often along cooling joints in the rock. It can also be found in hydrothermal mineral veins. It feels greasy when handled. Molybdenite is the chief ore of molybdenum and is mined in the USA, Russia, Canada and Chile. Molybdenum is used as a lubricant and has many electrical applications. It is also alloyed with other metals.

PYRITE

Group	Sulphides
Composition	FeS$_2$
Crystal System	Cubic
Habit	Crystals cubic, octahedral; massive, reniform, granular, botryoidal
Cleavage	Indistinct
Fracture	Uneven or conchoidal
Specific Gravity	5.0
Hardness	6 to 6½
Colour	Pale, silvery, yellow
Streak	Black
Lustre	Metallic

Pyrite forms in hydrothermal veins with other sulphides and in many rocks, including granite, slate and shale. Known by the vernacular name 'fool's gold' because of its colour, pyrite is common, much harder than gold, and has a much lower specific gravity. The crystals of pyrite frequently have striated faces and tarnish with iridescent colours when exposed to the air. Pyrite is the most common sulphide mineral.

Twinned, striated cubic crystals of pyrite with typical brassy-yellow colour and metallic lustre. Specimen from Cornwall, England.

ARSENOPYRITE

Group	Sulphides
Composition	FeAsS
Crystal System	Monoclinic
Habit	Crystals prismatic; columnar, massive
Cleavage	Indistinct
Fracture	Uneven
Specific Gravity	6.1
Hardness	5½ to 6
Colour	Silvery-white
Streak	Black
Lustre	Metallic

Arsenopyrite, the most abundant arsenic mineral, occurs in some igneous and metamorphic rocks, but forms mainly in hydrothermal veins. Its silvery-white colour readily tarnishes to iridescent coppery colours. The crystals often have striated faces. As with other minerals containing arsenic, a smell of garlic is produced when specimens are broken or heated. Much of the world's arsenic is obtained from arsenopyrite.

Massive and crystalline arsenopyrite. Some of the crystals are striated. Typical silvery colour and metallic lustre. Specimen from Portugal.

MARCASITE

Group	Sulphides
Composition	FeS_2
Crystal System	Orthorhombic
Habit	Crystals pyramidal, tabular; massive, nodular, reniform, spear-shaped
Cleavage	Distinct
Fracture	Uneven
Specific Gravity	4.9
Hardness	6 to 6½
Colour	Silvery-yellow, whitish
Streak	Greenish-black
Lustre	Metallic

Marcasite forms in sedimentary rocks such as clay, shale, chalk and limestone, often occurring as nodular lumps or flat, rounded masses with a radiating internal structure. It has the same chemical composition as pyrite, but is paler in colour. When exposed to the atmosphere, marcasite often darkens in colour and decomposes. It may also become iridescent when tarnished.

A flat marcasite specimen from southern England. This example has a flattened radiating structure.

TELLURIDES

Tellurides are compounds of metallic elements and tellurium.

SYLVANITE

Group	Tellurides
Composition	$AuAgTe_4$
Crystal System	Monoclinic
Habit	Crystals prismatic, bladed; columnar, granular
Cleavage	Perfect
Fracture	Uneven
Specific Gravity	8.2
Hardness	1½ to 2
Colour	Yellowish, grey, silvery-white
Streak	Silvery-white
Lustre	Metallic

Sylvanite occurs in hydrothermal veins, sometimes with native gold, and often with quartz, fluorite, sulphides including pyrite, carbonates such as rhodochrosite, and other tellurides. Sylvanite has a high specific gravity because it contains both gold and silver. If specimens remain in bright light, they become tarnished.

Granular sylvanite on rock matrix, showing grey colour and metallic lustre. Specimen from Ontario, Canada.

ARSENIDES

Arsenide minerals are compounds of metallic elements and arsenic.

NICKELINE

Group	Arsenides
Composition	NiAs
Crystal System	Trigonal/hexagonal
Habit	Crystals pyramidal; columnar, reniform, massive
Cleavage	None
Fracture	Uneven
Specific Gravity	7.8
Hardness	5 to 5½
Colour	Coppery-red
Streak	Black
Lustre	Metallic

Nickeline is found in hydrothermal mineral veins together with other nickel minerals and cobaltite and silver. The basic igneous rock norite may contain nickeline. A garlic smell is given off when the mineral is heated and a green solution is produced when nickeline is added to nitric acid. Nickel was first obtained from this mineral. Niccolite is an alternative name.

Massive nickeline with typical coppery colouring and metallic lustre. Specimen from Western Australia.

SKUTTERUDITE

Group	Arsenides
Composition	$(Co,Ni)As_3$
Crystal System	Cubic
Habit	Crystals cubic; granular, massive
Cleavage	Distinct
Fracture	Uneven
Specific Gravity	6.1 to 6.9
Hardness	5½ to 6
Colour	Tin-white
Streak	Black
Lustre	Metallic

Skutterudite is found in hydrothermal mineral veins, with a variety of other minerals, including native silver, arsenopyrite, erythrite, nickeline, cobaltite, siderite, barite, quartz and calcite. As is usual with minerals which contain arsenic, a smell of garlic is produced on heating. Skutterudite has been used as an ore of nickel and cobalt. When tarnished, this mineral becomes greyish or iridescent.

Tin-white skutterudite with massive habit and metallic lustre. The white mineral is calcite. Specimen from Morocco.

SULPHOSALTS

Sulphosalt minerals are composed of metals combined with sulphur and semi-metallic elements, often including arsenic and antimony.

ENARGITE

Group	Sulphosalts
Composition	Cu_3AsS_4
Crystal System	Orthorhombic
Habit	Crystals tabular, prismatic; granular, massive
Cleavage	Perfect
Fracture	Uneven
Specific Gravity	4.4
Hardness	3
Colour	Black
Streak	Black
Lustre	Metallic

Enargite occurs in hydrothermal mineral veins in association with other hydrothermal minerals including sphalerite and galena. This mineral is also found in replacement deposits with quartz and sulphides. It has been found in the insoluble cap rocks of salt domes. As with other minerals containing arsenic, a characteristic smell of garlic is produced when it is heated.

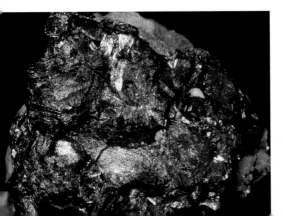

Black enargite with massive habit, good cleavage and metallic lustre. Specimen from Peru.

Black bournonite with granular habit and some small prismatic crystals. The transparent prismatic crystals are quartz. Specimen from Romania.

BOURNONITE

Group	Sulphosalts
Composition	$CuPbSbS_3$
Crystal System	Orthorhombic
Habit	Crystals tabular, prismatic; massive, granular
Cleavage	Imperfect
Fracture	Uneven or subconchoidal
Specific Gravity	5.8
Hardness	2½ to 3
Colour	Black or grey
Streak	Steel grey to black
Lustre	Metallic

Bournonite forms in hydrothermal mineral veins and is found with many other hydrothermal minerals, including galena, sphalerite, stibnite, siderite, quartz, chalcocite, tetrahedrite and chalcopyrite. It is soluble in nitric acid, copper in its composition turning the acid green. Its crystal faces are often striated. Bournonite may be altered by weathering or hydrothermal fluids to azurite, malachite or cerussite.

Black tetrahedrite crystals with tetrahedral habit and metallic lustre. The white prismatic crystals are quartz. Specimen from Peru.

TETRAHEDRITE

Group	Sulphosalts
Composition	$Cu_{12}Sb_4S_{13}$
Crystal System	Cubic
Habit	Crystals tetrahedral; granular, massive
Cleavage	None
Fracture	Subconchoidal or uneven
Specific Gravity	4.6 to 5.1
Hardness	3 to 4½
Colour	Black
Streak	Brown, black, reddish
Lustre	Metallic

Tetrahedrite forms with sulphides such as galena and sphalerite, and with barite, fluorite, quartz, silver and copper minerals, in hydrothermal mineral veins. It may also be found in very coarse-grained granitic pegmatites. Tetrahedrite dissolves in nitric acid. The mineral tennantite is closely related to tetrahedrite. Their chemistry differs in that tennantite contains arsenic instead of antimony. Tetrahedrite is the commoner of the two.

HALIDES

Halides are a group of minerals formed when metals combine with a halogen element such as iodine, chlorine, bromine or fluorine.

HALITE

Group	Halides
Composition	NaCl
Crystal System	Cubic
Habit	Crystals cubic, octahedral; massive, granular
Cleavage	Perfect
Fracture	Conchoidal or uneven
Specific Gravity	2.2
Hardness	2
Colour	Colourless, white, orange, brown, grey
Streak	White
Lustre	Vitreous or resinous

Halite forms mainly in evaporite deposits. These occur when saline water, such as an inland salt lake or a marine lagoon, dries out. Chemicals held in the water are precipitated as it evaporates and minerals including halite form, often interbedded with clay and marl. Layers of halite are known as rock salt. Halite is readily identified by its salty taste. This mineral (sodium chloride) is much used in the chemical industry. Some cubic halite crystals have stepped, concave faces and are called hopper crystals. Halite is soluble in water.

Massive halite, coloured by impurities, showing perfect cleavage and vitreous lustre. The rounded parts of the specimen indicate its solubility in water. Specimen from Cheshire, England.

SYLVITE

Group	Halides
Composition	KCl
Crystal System	Cubic
Habit	Crystals cubic, octahedral; granular, massive, encrusting
Cleavage	Perfect
Fracture	Uneven
Specific Gravity	2.0
Hardness	2
Colour	White, grey, reddish, yellow, colourless
Streak	White
Lustre	Vitreous

Sylvite forms in evaporite deposits with other evaporite minerals including halite, gypsum and polyhalite. These minerals are precipitated in a strict sequence as the saline water evaporates, the least soluble mineral, often gypsum, forming first and the most soluble last. Sylvite is an important industrial mineral, being mined as a source of agricultural fertiliser, and known as potash.

Twinned octahedral crystals of sylvite showing vitreous lustre and typical pale colouring. Specimen from North Yorkshire, England.

ATACAMITE

Group	Halides
Composition	$Cu_2Cl(OH)_3$
Crystal System	Orthorhombic
Habit	Crystals tabular, prismatic; granular, massive, fibrous
Cleavage	Perfect
Fracture	Conchoidal
Specific Gravity	3.8
Hardness	3 to 3½
Colour	Green
Streak	Pale green
Lustre	Vitreous or adamantine

Atacamite is usually found near copper deposits, especially where copper minerals have been altered by weathering and oxidation. It has also been discovered in fumaroles and the crusts around deep seabed black smokers. The copper minerals azurite and malachite often occur with it, as may cuprite, brochantite, linarite and chrysocolla. Atacamite dissolves in hydrochloric acid, but no effervescence takes place.

Delicate dark green, prismatic atacamite crystals with vitreous lustre. Specimen from Chile.

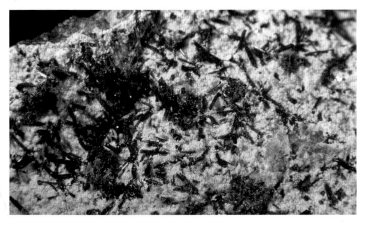

FLUORITE

Group	Halides
Composition	CaF_2
Crystal System	Cubic
Habit	Crystals cubic, octahedral; granular, massive
Cleavage	Perfect
Fracture	Conchoidal
Specific Gravity	3.2
Hardness	4
Colour	Green, blue, purple, white, yellowish, red
Streak	White
Lustre	Vitreous

Fluorite forms in hydrothermal mineral veins and in the fractures in granites. It also occurs around hot springs. In hydrothermal veins it is found with many minerals including quartz, calcite, galena, barite, chalcopyrite, pyrite and sphalerite. Fluorite is mined in China, Mexico and Europe. It is used as a flux in steel making and is refined for fluorine, which is used in the manufacture of hydrofluoric acid and fluorine gas. Fluorocarbons have a wide

Purple cubes of fluorite with silvery-grey galena from County Durham, England.

Mass of small twinned fluorite cubes on limestone matrix. Specimen from County Durham, England.

range of uses as refrigerants, cleaning fluids and lubricants, but there are certain environmental concerns over their use. When fluorite cleaves, the corners of its cubic crystals break off to produce triangular surfaces. Some crystals are varicoloured, as in the banded form known as Blue John, which is used ornamentally. Fluorite is fluorescent under ultraviolet light.

Massive sectioned fluorite with colour-banded structure, edged with pyrite. Specimen from China.

Mass of minute diaboleite crystals with typical colouring. Specimen from Chile.

DIABOLEITE

Group	Halides
Composition	$Pb_2CuCl_2(OH)_4$
Crystal System	Tetragonal
Habit	Crystals tabular; plates, massive, granular
Cleavage	Perfect
Fracture	Conchoidal
Specific Gravity	5.4
Hardness	2½
Colour	Rich dark blue
Streak	Pale blue
Lustre	Vitreous

Diaboleite occurs when previously formed minerals are changed by weathering, or by hydrothermal fluids seeping from depth. It may be found with other lead- and copper-bearing minerals, including linarite, cerussite, atacamite and boleite. It also occurs with manganese ores and in slag which has been exposed to seawater.

OXIDES AND HYDROXIDES

Minerals in the oxide group are compounds of various elements combined with oxygen. Hydroxide minerals are compounds of metallic elements and the hydroxyl radical, (OH).

SPINEL

Group	Oxides
Composition	$MgAl_2O_4$
Crystal System	Cubic
Habit	Crystals cubic, dodecahedral; granular, massive
Cleavage	Indistinct
Fracture	Uneven or conchoidal
Specific Gravity	3.6
Hardness	7½ to 8
Colour	Brown, red, black, green, blue
Streak	White
Lustre	Vitreous

Spinel occurs in a number of metamorphic rocks including the contact metamorphic rock marble and the regionally metamorphosed gneiss. It has also been found in serpentinite and peridotite. Because of its hardness, spinel accumulates in alluvial sediments. There are two varieties of spinel with different chemistry: pleonaste contains iron and is dark in colour, while picotite contains chromium. Good quality spinel is used as a gemstone.

Small water-worn crystals of spinel with different colouring. Some, where broken, show the vitreous lustre. These examples are from a placer deposit in India.

ZINCITE

Group	Oxides
Composition	(Zn,Mn)O
Crystal System	Trigonal/hexagonal
Habit	Crystals pyramidal; massive, foliated, granular
Cleavage	Perfect
Fracture	Conchoidal
Specific Gravity	5.7
Hardness	4
Colour	Red, orange
Streak	Orange-yellow
Lustre	Adamantine

Zincite, a relatively rare mineral, is found mainly in metamorphic rocks formed by heating due to the proximity of igneous rocks. Willemite and calcite are associated minerals. Zincite can be manufactured synthetically and both natural and man-made zincite have been used as semiconductors.

Typical red zincite with dark franklinite from New Jersey, USA.

CUPRITE

Group	Oxides
Composition	Cu_2O
Crystal System	Cubic
Habit	Crystals cubic, octahedral; granular, massive, earthy
Cleavage	Poor
Fracture	Uneven or conchoidal
Specific Gravity	6.1
Hardness	3½ to 4
Colour	Bright red
Streak	Brown
Lustre	Metallic, adamantine or earthy

Cuprite forms where copper minerals and veins have been altered by oxidation. Native copper, iron oxides, azurite, malachite and chrysocolla are found with it. Because of its colour and lustre, cuprite has been used as a gemstone. That it has not become very valuable in this respect is due to its low hardness, as it is easily scratched in everyday use.

Massive and crystalline cuprite with typical red colouring. Two crystals at the bottom of the specimen show an adamantine lustre. Specimen from Chile.

Prismatic crystals of pyrolusite with characteristic metallic lustre.

PYROLUSITE

Group	Oxides
Composition	MnO_2
Crystal System	Tetragonal
Habit	Crystals prismatic; fibrous, dendritic, massive, columnar
Cleavage	Perfect
Fracture	Uneven
Specific Gravity	5.1
Hardness	2 to 6½
Colour	Black or dark grey
Streak	Black or bluish-black
Lustre	Metallic or dull

Pyrolusite occurs where manganese veins, containing manganite, have been altered, and in bogs and lakes. The manganese nodules found on the deep ocean bed contain pyrolusite. Some specimens are relatively hard, but the dendritic material is soft and leaves a sooty impression on the fingers. Dendritic pyrolusite is often mistaken for plant fossils, especially by the producers of flooring slabs, where their products are called 'fossil'. It has industrial use, including the green and violet colouring of paints, glass and pottery.

Tiny black crystals of magnetite, with typical metallic lustre, set in white calcite. Specimen from Norway.

MAGNETITE

Group	Oxides
Composition	Fe_3O_4
Crystal System	Cubic
Habit	Crystals octahedral, dodecahedral; granular, massive
Cleavage	None
Fracture	Uneven or subconchoidal
Specific Gravity	5.2
Hardness	5½ to 6½
Colour	Black
Streak	Black
Lustre	Metallic or dull

Magnetite occurs in mineral veins and as a replacement deposit. It also forms in basic igneous rocks and is found in beach and alluvial sands, where it accumulates because of its hardness and resistance to weathering. This mineral is an important ore of iron. Major sources include China, Peru, Sweden, Western Australia, the USA, South Africa and India. As its name suggests, magnetite is magnetic; a specimen will attract iron filings and move a compass needle.

A massive specimen of dark ilmenite showing a bright metallic lustre. Specimen from South Africa.

ILMENITE

Group	Oxides
Composition	$FeTiO_3$
Crystal System	Trigonal/hexagonal
Habit	Crystals rhombohedral, tabular; massive, granular, compact
Cleavage	None
Fracture	Uneven or conchoidal
Specific Gravity	4.7
Hardness	5 to 6
Colour	Brownish, black, silvery
Streak	Black
Lustre	Metallic or dull

Ilmenite occurs in small amounts in many igneous rocks, such as coarse-grained pegmatites, and also in hydrothermal mineral veins. It accumulates in alluvial sands as a placer mineral. Ilmenite is the main ore of titanium and the mineral is found in huge amounts in Western Australia in sand deposits. Economically useful quantities also exist in Canada, Norway and the Ukraine. Titanium is used in making aircraft engines and framework, because it has the strength of steel but is 45% lighter and does not suffer metal fatigue.

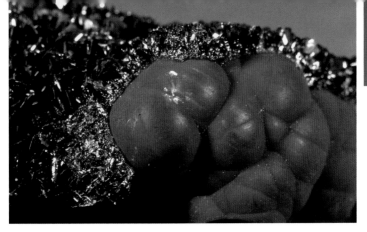

This specimen shows reddish, reniform kidney ore and black crystalline specularite. Specimen from Cumbria, England.

HAEMATITE

Group	Oxides
Composition	Fe_2O_3
Crystal System	Trigonal/hexagonal
Habit	Crystals prismatic, rhombohedral, tabular; reniform, massive, fibrous
Cleavage	None
Fracture	Uneven or subconchoidal
Specific Gravity	5.3
Hardness	5 to 6
Colour	Brown, reddish, black
Streak	Red-brown
Lustre	Metallic or dull

Haematite forms as a replacement mineral in sedimentary rock, especially limestone, and in hydrothermal veins. It commonly occurs in a reniform habit with a reddish colour, giving rise to its name of kidney ore. Haematite is an important metal ore, having a high iron content. Large deposits are worked in Canada, Brazil and Venezuela. The black crystalline variety of haematite is known as specularite. A form called iron rose is composed of small tabular crystals in a rosette.

CASSITERITE

Group	Oxides
Composition	SnO_2
Crystal System	Tetragonal
Habit	Crystals prismatic, pyramidal; reniform, massive, granular, botryoidal
Cleavage	Imperfect
Fracture	Uneven or subconchoidal
Specific Gravity	7.0
Hardness	6 to 7
Colour	Black, brown, yellow
Streak	Brown, grey
Lustre	Greasy or adamantine

Cassiterite occurs in hydrothermal mineral veins, with many other minerals including pyrite, chalcopyrite, quartz, sphalerite and galena. This mineral is the main ore of tin. It is mined in Malaysia, Burma, China and Thailand, and in Brazil and Bolivia. The presence of cassiterite in Cornwall, England, was exploited for hundreds of years until the world market changed and the industry there faded. The main uses of tin are in alloys and for tin plating.

Black crystals of cassiterite with adamantine lustre from Bolivia.

CHRYSOBERYL

Group	Oxides
Composition	$BeAl_2O_4$
Crystal System	Orthorhombic
Habit	Crystals prismatic, tabular; massive, granular
Cleavage	Distinct
Fracture	Uneven or conchoidal
Specific Gravity	3.7
Hardness	8½
Colour	Yellow, greenish
Streak	White
Lustre	Vitreous

Chrysoberyl occurs in coarse-grained igneous pegmatites, as well as metamorphic marbles, schists and gneisses. Because of its great hardness and resistance to physical weathering, chrysoberyl is found in alluvial sands. With a considerable hardness, vitreous lustre and transparency, chrysoberyl has a gem variety, alexandrite. This has the remarkable property of being red in artificial light and green in daylight.

Pale greenish crystals and grains of chrysoberyl set in white quartz.

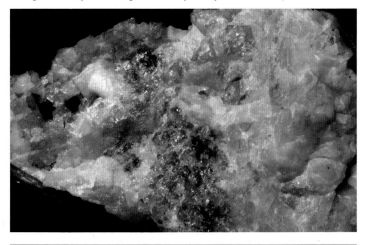

CORUNDUM

Group	Oxides
Composition	Al_2O_3
Crystal System	Trigonal/hexagonal
Habit	Crystals rhombohedral, tabular, prismatic, bipyramidal; massive, granular
Cleavage	None
Fracture	Uneven or conchoidal
Specific Gravity	4.0
Hardness	9
Colour	Brownish, grey, blue, red, yellowish, white
Streak	White
Lustre	Adamantine or vitreous

Corundum occurs in basic igneous rocks and in some metamorphic rocks. It is one of the very hardest minerals, being the defining mineral at point nine on the hardness scale. Corundum is used as an abrasive, but the coloured forms are greatly valued as gemstones. Sapphire (blue, illustrated on p.75) and ruby (red) are colour varieties of corundum. Both can be found in alluvial placer deposits, where they accumulate because of their great hardness.

Red corundum (ruby) crystals set in calcite. Specimen from India.

Brownish, prismatic corundum crystal from Zimbabwe.

RUTILE

Group	Oxides
Composition	TiO_2
Crystal System	Tetragonal
Habit	Crystals acicular, prismatic; massive
Cleavage	Distinct
Fracture	Uneven or conchoidal
Specific Gravity	4.2
Hardness	6 to 6½
Colour	Black, brown, red, yellowish
Streak	Yellow, brown
Lustre	Adamantine or metallic

Rutile commonly occurs as acicular (needle-shaped) crystals within quartz, and as granular or compact masses. It is also found as an accessory mineral in many igneous rocks and in metamorphic rocks such as gneiss, metaquartzite and schist. Along with anatase and brookite, which have the same chemical formula as rutile and form a series of minerals with it, rutile is an important ore of titanium.

Slender, needle-like crystals of rutile, with acicular habit, embedded in transparent, colourless quartz.

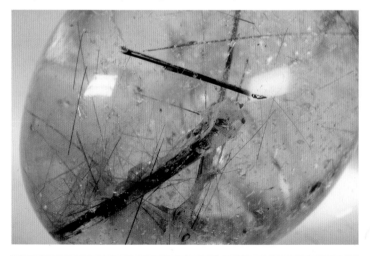

Typical dark uraninite, with uneven fracture and submetallic lustre.

URANINITE

Group	Oxides
Composition	UO_2
Crystal System	Cubic
Habit	Crystals cubic, octahedral, dodecahedral; massive, granular, botryoidal
Cleavage	Indistinct
Fracture	Uneven or conchoidal
Specific Gravity	6.5 to 10
Hardness	5 to 6
Colour	Dark brown, black
Streak	Black, brown
Lustre	Greasy, dull, pitch-like or submetallic

Uraninite occurs in hydrothermal mineral veins and in acid igneous rocks, including granite and pegmatite. It can also be found in some sandstones. Uraninite is a radioactive mineral and great care and sensible precautions have to be taken when examining a specimen. An alternative name, pitchblende, arises from the pitch-like colour of some forms of the mineral. It is the main ore of uranium, and is mined in the USA, Australia, Canada, China, Russia and South Africa.

QUARTZ

Group	Oxides
Composition	SiO_2
Crystal System	Trigonal/hexagonal
Habit	Crystals prismatic, pyramidal, rhombohedral; massive, granular
Cleavage	None
Fracture	Uneven or conchoidal
Specific Gravity	2.65
Hardness	7
Colour	White, colourless, grey, pink, purple, black, brown, yellow, green
Streak	White
Lustre	Vitreous

Quartz is one of the most common minerals. It is often classified with the silicate group of minerals, but strictly it is an oxide. Quartz forms in hydrothermal mineral veins and in many rocks, especially granite, gneiss, schist and metaquartzite. In mineral veins it occurs with metallic ore minerals and is referred to as a 'gangue' mineral. Many of the colour varieties have their own names and

A mass of small, transparent, hexagonal quartz prisms, with grey, metallic galena.

Purple amethyst from a geode. Specimen from Brazil.

are used as gemstones. These include purple amethyst, pink rose quartz and yellow citrine. Quartz has many economic uses and when found in great bulk, as in sand and sandstone, is used in the construction industry. Much of the silicon used in the electronics industry is made artificially. Quartz defines point seven on the hardness scale.

Dark Cairngorm (smoky quartz) specimens. One shows hexagonal prisms, another is polished and two are cut and polished.

A mass of botryoidal chalcedony coated with minute sparkling quartz crystals.

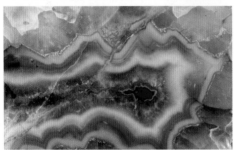

Agate showing alternating bands of red, white and blue chalcedony. Specimen from the Cheviot Hills, Scottish Borders.

CHALCEDONY

Group	Oxides
Composition	SiO_2
Crystal System	Trigonal/hexagonal
Habit	Microcrystalline; massive, botryoidal or mammilated
Cleavage	None
Fracture	Conchoidal or uneven
Specific Gravity	2.65
Hardness	7
Colour	White, brown, green, blue, pink, red, black
Streak	White
Lustre	Vitreous or waxy

Chalcedony is found in veins and cavities in a variety of rocks, especially in cavities in igneous rocks. It has a similar chemical composition to quartz, but is microcrystalline. The colour varieties have different names and many are used as ornamental stones and gemstones. The red variety is carnelian, chrysoprase is green and sard is light to dark brown. Agate is chalcedony with a concentric banded structure, while jasper is red and opaque.

Shimmering opal from Australia, showing vitreous lustre and varied colouring.

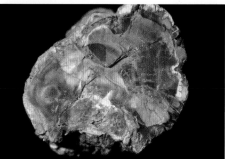

Brownish wood opal with some concentric rings. Specimen from Arizona, USA.

OPAL

Group	Oxides
Composition	$SiO_2 \cdot nH_2O$
Crystal System	Amorphous
Habit	Non-crystalline; massive, concretionary, botryoidal, globular, reniform
Cleavage	None
Fracture	Conchoidal
Specific Gravity	2.0 to 2.2
Hardness	5½ to 6½
Colour	White, blue, yellow, red, green, black
Streak	White
Lustre	Waxy, resinous or vitreous

Opal occurs as the result of precipitation from silica-rich solutions, especially around hot springs. It is used as a gemstone and has a very rich play of colours which can change when the mineral is warmed. The red to orange variety is known as fire opal. Wood opal is the form which replaces plant material and can have concentric tree-rings preserved in its structure. Hyalite is colourless, with a bubbly habit. If heated, opal loses its water molecules and may become quartz.

GOETHITE

Group	Hydroxides
Composition	FeO(OH)
Crystal System	Orthorhombic
Habit	Crystals prismatic; massive, stalactitic, botryoidal
Cleavage	Perfect
Fracture	Uneven
Specific Gravity	3.3 to 4.3
Hardness	5 to 5½
Colour	Yellow-brown, brown, black
Streak	Yellow-brown
Lustre	Dull or adamantine

Goethite occurs in regions where iron minerals have been altered by oxidation. This can be the result of weathering or the action of subsurface fluids. Though not as rich in iron as some ores, goethite is used as an ore of this metal. Limonite, usually yellowish-brown in colour, is a similar mineral, but with an added molecule of water.

A mass of small, dark, prismatic goethite crystals. Specimen from Germany.

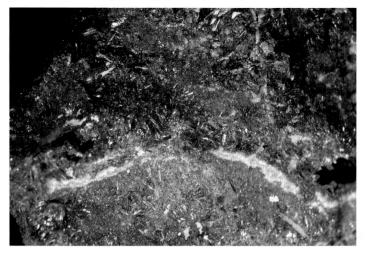

A massive specimen of manganite from Sweden showing submetallic lustre and dark colouring.

MANGANITE

Group	Hydroxides
Composition	$MnO(OH)$
Crystal System	Monoclinic
Habit	Crystals prismatic; massive, columnar, granular, fibrous, concretionary
Cleavage	Perfect
Fracture	Uneven
Specific Gravity	4.3
Hardness	4
Colour	Black, dark grey
Streak	Black or red-brown
Lustre	Submetallic

Manganite occurs in hydrothermal mineral veins and in bogs and lakes. It is often found in clay and laterite deposits with other oxides of manganese. Crystalline manganite frequently forms as bundles of striated prismatic crystals. It can be altered to pyrolusite by fluids circulating through the Earth's crust. Manganite has been used as an ore of manganese.

CARBONATES

Carbonates are compounds of metal or semi-metal elements with the carbonate radical (CO_3).

ARAGONITE

Group	Carbonates
Composition	$CaCO_3$
Crystal System	Orthorhombic
Habit	Crystals prismatic, pseudo-hexagonal; radiating, fibrous, coralline
Cleavage	Distinct
Fracture	Subconchoidal
Specific Gravity	2.9
Hardness	3½ to 4
Colour	White, colourless, grey, blue, green, red, brown
Streak	White
Lustre	Vitreous

Aragonite occurs around hot springs and in cave deposits, especially in limestone regions. It is also found in hydrothermal veins. The coral-like form is called 'flos-ferri'. Aragonite dissolves readily in dilute hydrochloric acid, with effervescence. This mineral is also produced biologically in mollusc shells, and has the same chemical formula as calcite.

Prismatic aragonite crystal with a hexagonal habit and smaller crystals on its surface. Specimen from Spain.

Prismatic brownish aragonite crystals, with a hexagonal cross-section.

LEFT: *Pale prismatic calcite crystals, some with double terminations. Specimen from Cumbria, England.* ABOVE: *Nail-head calcite crystals showing the typical terminations which give the crystals this name.*

CALCITE

Group	Carbonates
Composition	$CaCO_3$
Crystal System	Trigonal/hexagonal
Habit	Crystals rhombohedral, scalenohedral, prismatic; massive, stalactitic, fibrous
Cleavage	Perfect
Fracture	Conchoidal
Specific Gravity	2.7
Hardness	3
Colour	White, colourless, brownish, grey, reddish
Streak	White
Lustre	Vitreous or pearly

Calcite is common in hydrothermal mineral veins, limestones and marbles. Some sandstones have a calcareous cement composed mainly of calcite. This mineral defines point three on the hardness scale and dissolves with effervescence in weak hydrochloric acid. Calcite cleaves into rhombs. Some crystals have distinctive ends and are called 'nail-head' crystals, while others with sharp terminations are 'dog-tooth' crystals. Rhombs of calcite (Iceland spar) exhibit double refraction; objects seen through the rhomb appear double, as shown on p.75.

RHODOCHROSITE

Group	Carbonates
Composition	$MnCO_3$
Crystal System	Trigonal/hexagonal
Habit	Crystals prismatic, tabular; massive, botryoidal, stalactitic
Cleavage	Perfect
Fracture	Uneven or conchoidal
Specific Gravity	3.7
Hardness	3½ to 4
Colour	Red, pink, white, brown, orange
Streak	White
Lustre	Vitreous or pearly

The alteration of other manganese minerals can produce rhodochrosite. It also occurs in hydrothermal veins. Rhodochrosite is rather soft for use as a gemstone, but because of its colour and frequently banded structure it is often cut and polished ornamentally. Rhodochrosite is a source of manganese which is alloyed with steel to increase the strength of this metal. It dissolves with effervescence in weak hydrochloric acid.

Deep red crystals of vitreous rhodochrosite on quartz-rich matrix. Specimen from Argentina.

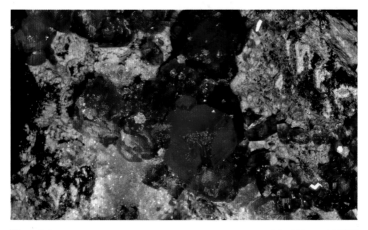

DOLOMITE

Group	Carbonates
Composition	$CaMg(CO_3)_2$
Crystal System	Trigonal/hexagonal
Habit	Crystals rhombohedral; granular, massive
Cleavage	Perfect
Fracture	Subconchoidal
Specific Gravity	2.8
Hardness	3½ to 4
Colour	White, colourless, brownish, grey
Streak	White
Lustre	Pearly or vitreous

Dolomite commonly occurs in hydrothermal mineral veins and in certain limestones which, because of the abundance of this mineral in them, are referred to as dolomitic limestone or dolostone. The small rhombohedral crystals which are typical of dolomite frequently have curved crystal faces and this can be a good identification feature. These faces often form a 'saddle-shaped' crystal. Dolomite dissolves very slowly in dilute hydrochloric acid.

A mass of small brownish dolomite crystals showing typical curved faces. Specimen from Spain.

SMITHSONITE

Group	Carbonates
Composition	$ZnCO_3$
Crystal System	Trigonal/hexagonal
Habit	Crystals rhombohedral; reniform, botryoidal, massive, granular
Cleavage	Perfect
Fracture	Uneven or subconchoidal
Specific Gravity	4.3 to 4.4
Hardness	4 to 4½
Colour	Blue, green, grey, yellow, purple, brown, white
Streak	White
Lustre	Vitreous or pearly

Smithsonite is associated with the parts of copper and zinc veins which have been altered and oxidised, often by weathering. Azurite and malachite are two brightly coloured minerals which occur with smithsonite, as well as aurichalcite, willemite, cerussite and anglesite. In common with other carbonate minerals, it is soluble in dilute hydrochloric acid. Smithsonite may have curved crystal faces.

Botryoidal smithsonite with pale yellowish-green colouring. Specimen from Ireland.

SIDERITE

Group	Carbonates
Composition	$FeCO_3$
Crystal System	Trigonal/hexagonal
Habit	Prismatic, rhombohedral, tabular; massive, botryoidal, granular
Cleavage	Perfect
Fracture	Uneven or conchoidal
Specific Gravity	4.0
Hardness	4
Colour	Brown, yellowish, green, grey, reddish
Streak	White
Lustre	Vitreous

Siderite occurs in hydrothermal mineral veins with many other minerals including quartz, calcite, galena, sphalerite and barite. It is also found in some sedimentary rocks with an oolitic texture. The rhombohedral crystals often have curved faces like those of dolomite. Siderite is an ore of iron and is free from sulphur and phosphorus, which other ores may contain.

Small twinned crystals of pale brown siderite with transparent quartz. Specimen from Peru.

MAGNESITE

Group	Carbonates
Composition	$MgCO_3$
Crystal System	Trigonal/hexagonal
Habit	Crystals tabular, rhombohedral, prismatic; massive, granular, fibrous
Cleavage	Perfect
Fracture	Uneven or conchoidal
Specific Gravity	3.1
Hardness	3 to 4
Colour	White, colourless, brown, grey, yellow
Streak	White
Lustre	Dull or vitreous

Magnesite occurs when ultrabasic rocks and serpentinites are altered. It can also be found in regionally and thermally metamorphosed rocks and as a gangue mineral in hydrothermal veins. Crystals are uncommon; magnesite is usually massive. When magnesium-rich rocks are altered by weathering and fluids underground, magnesite can be formed. It is used as a flux in steel production and in the manufacture of synthetic rubber.

Massive grey magnesite with dull lustre.

CERUSSITE

Group	Carbonates
Composition	$PbCO_3$
Crystal System	Orthorhombic
Habit	Crystals bladed, tabular; massive, granular
Cleavage	Distinct
Fracture	Conchoidal
Specific Gravity	6.6
Hardness	3 to 3½
Colour	Brownish, grey, white, colourless, green
Streak	White
Lustre	Vitreous or adamantine

Cerussite occurs in the altered parts of mineral veins that contain lead minerals, especially galena (lead sulphide). The crystals are frequently twinned. The presence of lead in the composition of cerussite gives the mineral a high specific gravity, a good identification feature for a pale-coloured mineral.

A mass of bladed cerussite crystals showing typical greyish and brownish colouring with a vitreous lustre. This variety is often called 'Jack straw'. The specimen is from County Durham, England.

MALACHITE

Group	Carbonates
Composition	$Cu_2CO_3(OH)_2$
Crystal System	Monoclinic
Habit	Crystals prismatic, acicular; botryoidal, stalactitic
Cleavage	Perfect
Fracture	Uneven or subconchoidal
Specific Gravity	4.0
Hardness	3½ to 4
Colour	Bright green
Streak	Green
Lustre	Vitreous or silky

Malachite is found in altered copper zones, frequently with azurite. The formation of malachite is often the result of chemical reactions between copper-bearing sulphide minerals and carbonate minerals. Malachite is well known from major copper-producing regions and is used as a decorative stone because of its striking green colour and banded internal structure. As it is very soft, malachite takes a high polish and is easily worked.

Rich green, botryoidal malachite, with small amounts of azurite, from Congo, Africa.

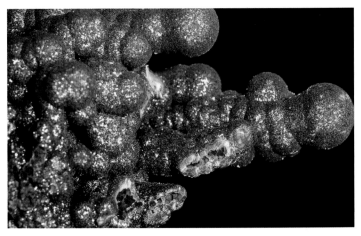

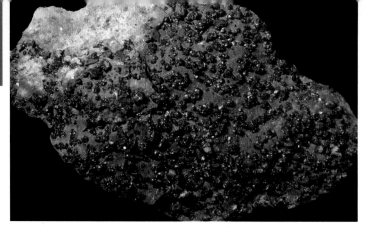

Tiny deep blue azurite crystals set on a matrix of limonite-rich rock. Specimen from Morocco.

AZURITE

Group	Carbonates
Composition	$Cu_3(CO_3)_2(OH)_2$
Crystal System	Monoclinic
Habit	Crystals prismatic, tabular; stalactitic, nodular, botryoidal
Cleavage	Perfect
Fracture	Uneven or conchoidal
Specific Gravity	3.8
Hardness	3½ to 4
Colour	Deep blue
Streak	Pale blue
Lustre	Vitreous or dull

Azurite occurs with malachite in the altered zones of copper veins, which may also contain calcite, chrysocolla and chalcocite. Azurite is less stable than malachite and can be altered to and replaced by malachite during weathering and the action of underground fluids. Because of its striking colour, azurite is sometimes used as an ornamental stone. It is soft and easily carved and takes a good polish.

BORATES

Borates are formed when metallic elements combine with the borate radical (BO_3). They are uncommon minerals.

ULEXITE

Group	Borates
Composition	$NaCaB_5O_6(OH)_6.5H_2O$
Crystal System	Triclinic
Habit	Crystals acicular, prismatic; tufted, fibrous
Cleavage	Perfect
Fracture	Uneven
Specific Gravity	2.0
Hardness	2½
Colour	Colourless, white
Streak	White
Lustre	Silky or vitreous

Ulexite occurs as an evaporite mineral, found with other evaporites such as halite and gypsum and also with calcite, trona, borax and colemanite. It commonly forms as tufts of acicular crystals. Ulexite has an unusual property which gives it the name of 'television stone': the fibrous crystals can transmit light by internal reflection, acting as optical fibres. If a specimen is cut and polished at each end, light is readily transmitted along the crystal.

A prismatic crystal of ulexite with a silky lustre from California, USA.

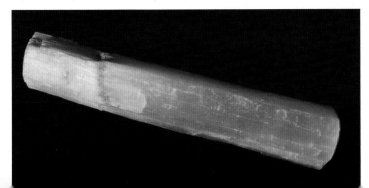

SULPHATES

Sulphate minerals have the chemical composition of a metallic element combined with the sulphate radical (SO_4).

GYPSUM

Group	Sulphates
Composition	$CaSO_4.2H_2O$
Crystal System	Monoclinic
Habit	Crystals diamond-shaped, tabular; fibrous, massive, rosettes
Cleavage	Perfect
Fracture	Splintery
Specific Gravity	2.3
Hardness	2
Colour	White, colourless, brown, reddish, grey, yellow, green
Streak	White
Lustre	Pearly, vitreous or dull

Gypsum forms in evaporite basins and on the margins of hot springs, with halite and sylvite. It also occurs in clay and shale. Crystals with high transparency are called selenite. Satin spar is the fibrous variety. Gypsum defines point two on the hardness scale and can readily be scratched with a fingernail. This mineral has important industrial uses and is mined for the production of plaster and plasterboard. In the caves of the Naica Mine at Chihuahua in Mexico, giant gypsum crystals up to 11 metres long, weighing 55 tonnes, have been found.

Diamond shaped gypsum crystals with a specimen of fibrous satin spar.

ABOVE: A mass of radiating clusters of transparent, prismatic crystals of gypsum, variety selenite. Specimen from Morocco. *BELOW:* Gypsum, variety desert rose, showing flattened crystals coated with small sand grains.

CELESTITE

Group	Sulphates
Composition	$SrSO_4$
Crystal System	Orthorhombic
Habit	Crystals prismatic, tabular, pyramidal; nodular, granular, fibrous
Cleavage	Perfect
Fracture	Uneven
Specific Gravity	4.0
Hardness	3 to 3½
Colour	White, yellow, grey, brown, reddish, green, blue
Streak	White
Lustre	Vitreous

Celestite occurs in hydrothermal veins and cavities in lavas. These are formed when gas bubbles are left as cavities in solidified lava, and when filled with crystals these cavities are known as geodes. Celestite is also found in some limestones and evaporite deposits. It is an important source of strontium, which is used in glass making and in fireworks. An alternative name for this mineral is celestine.

Bluish, transparent pyramidal crystals of celestite, some showing twinning. Specimen from Poland.

ANHYDRITE

Group	Sulphates
Composition	$CaSO_4$
Crystal System	Orthorhombic
Habit	Crystals prismatic, tabular, bladed; fibrous, massive, granular
Cleavage	Perfect
Fracture	Splintery or uneven
Specific Gravity	3.0
Hardness	3½
Colour	Grey, white, red, pink, blue
Streak	White
Lustre	Pearly, greasy or vitreous

Anhydrite occurs in evaporite deposits with gypsum, halite and sylvite. It can also be found in salt domes, where it acts as part of the cap rock along with halite. It is closely related to gypsum, the chemistry of the two minerals being very similar; gypsum has two water molecules, which anhydrite lacks. Anhydrite has the same uses as gypsum, in plaster and as a fertiliser.

Bladed prismatic crystals of pale blue anhydrite with a vitreous lustre.

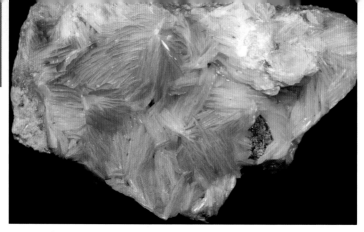

A mass of white and yellowish-brown barite crystals from County Durham, England.

BARITE

Group	Sulphates
Composition	$BaSO_4$
Crystal System	Orthorhombic
Habit	Crystals prismatic, tabular; columnar, massive, fibrous, lamellar
Cleavage	Perfect
Fracture	Uneven
Specific Gravity	4.5
Hardness	3 to 3½
Colour	Colourless, white, reddish, yellow, brown, grey, blue
Streak	White
Lustre	Pearly, resinous or vitreous

Barite commonly forms in hydrothermal mineral veins with many minerals such as quartz, calcite, sphalerite, galena, chalcopyrite and pyrite. It can also occur around hot springs and in nodules in clay. It is noticeably heavy for a common, pale-coloured mineral because of the presence of barium in its composition. Barite is the main ore of barium, which is used in diagnostic X-rays and as drilling mud in oil and gas exploration. A form called 'desert rose' occurs when rosettes of tabular crystals combine with sand grains. This mineral is sometimes called barytes.

LINARITE

Group	Sulphates
Composition	$PbCu(SO_4)(OH)_2$
Crystal System	Monoclinic
Habit	Crystals prismatic, tabular; crusts
Cleavage	Perfect
Fracture	Conchoidal
Specific Gravity	5.3
Hardness	2½
Colour	Rich blue
Streak	Pale blue
Lustre	Vitreous

Linarite is found in the parts of copper deposits that have been altered and oxidised. It occurs with other secondary minerals, such as chalcanthite and anglesite. It is a very soft mineral, but because of the lead and copper in its composition it has a high specific gravity. This and the colour are good identification features. Though similar to azurite, linarite does not react with dilute hydrochloric acid.

A crust of minute blue linarite crystals on quartz-rich rock. Specimen from Wales.

CHROMATES

Chromate minerals are compounds of metals and the chromate radical (CrO_4).

CROCOITE

Group	Chromates
Composition	$PbCrO_4$
Crystal System	Monoclinic
Habit	Crystals prismatic; massive
Cleavage	Distinct
Fracture	Uneven or conchoidal
Specific Gravity	6.0
Hardness	2½ to 3
Colour	Red, yellow, orange
Streak	Orange-yellow
Lustre	Adamantine or vitreous

Crocoite is associated with chromium and lead deposits and is formed when these are oxidised. It occurs with vanadinite, pyromorphite and cerussite. The prismatic crystals are usually twinned and occur in aggregates. Crocoite is a very striking mineral, with a high specific gravity because of the lead content.

A mass of slender prismatic crocoite crystals with typical orange-red colouring. Specimen from Tasmania.

MOLYBDATES

Minerals in this group are compounds of metals and the molybdate radical (MoO_4).

WULFENITE

Group	Molybdates
Composition	$PbMoO_4$
Crystal System	Tetragonal
Habit	Crystals prismatic, tabular; massive, granular
Cleavage	Distinct
Fracture	Subconchoidal or uneven
Specific Gravity	6.5 to 7
Hardness	2½ to 3
Colour	Yellow, orange, brown, greenish, grey
Streak	White
Lustre	Vitreous or resinous

Wulfenite is found where mineral veins containing lead ores, such as galena, have been changed by subsurface fluids. A great many vein minerals can occur with wulfenite, including galena, cerussite, pyromorphite, anglesite, hemimorphite, vanadinite and mimetite. This is another mineral with a high specific gravity because of its lead content. The tabular crystals frequently have a square outline.

Thin tabular wulfenite crystals, some showing twinning, with characteristic yellowish colouring and vitreous lustre.

PHOSPHATES

Phosphates are chemical compounds of metals and the phosphate radical (PO_4).

PYROMORPHITE

Group	Phosphates
Composition	$Pb_5(PO_4)_3Cl$
Crystal System	Trigonal/hexagonal
Habit	Crystals barrel-shaped, prismatic; botryoidal, fibrous, reniform, globular
Cleavage	Poor
Fracture	Uneven or subconchoidal
Specific Gravity	7.0
Hardness	3½ to 4
Colour	Yellowish, green, grey, brown, orange
Streak	White
Lustre	Adamantine or resinous

Pyromorphite occurs in the zones of lead-bearing veins that have been altered. It is found with galena and cerussite. Though forming as a secondary mineral, it has been mined for its lead content. The greenish-coloured, barrel-shaped crystals are fairly typical, together with the high specific gravity. Pyromorphite and mimetite are so similar physically that it is sometimes only possible to distinguish between them by chemical analysis.

Small, rounded, barrel-shaped crystals of pyromorphite showing resinous lustre. Specimen from Cumbria, England.

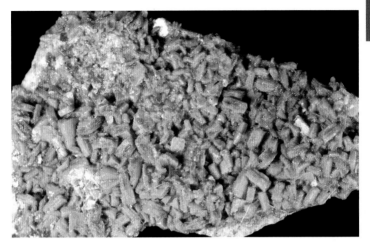

Tabular, greenish-yellow autunite crystals. Specimen from Washington, USA.

AUTUNITE

Group	Phosphates
Composition	$Ca(UO_2)_2(PO_4)_2 \cdot 10–12H_2O$
Crystal System	Tetragonal
Habit	Crystals tabular; granular, crusts
Cleavage	Perfect
Fracture	Uneven
Specific Gravity	3.0 to 3.2
Hardness	2 to 2½
Colour	Bright yellow, greenish
Streak	Pale yellow
Lustre	Vitreous or pearly

Autunite is a secondary mineral, occurring when minerals containing uranium are altered. It is found in granites and pegmatites with torbernite and uraninite, from which it forms. Autunite is recognised by its bright yellow or green colouring and short tabular crystals. It is a radioactive mineral and great care should be taken when examining specimens.

TORBERNITE

Group	Phosphates
Composition	$Cu(UO_2)_2(PO_4)_2 \cdot 8{-}12H_2O$
Crystal System	Tetragonal
Habit	Crystals tabular; lamellar
Cleavage	Perfect
Fracture	Uneven
Specific Gravity	3.2
Hardness	2 to 2½
Colour	Green
Streak	Green
Lustre	Pearly or vitreous

Torbernite forms by the alteration of uranium minerals such as uraninite. It occurs with autunite in these alteration zones. It is very similar to autunite, but the two minerals differ in colour and chemical composition, torbernite containing copper and autunite calcium. As with autunite, great care must be taken when dealing with this radioactive mineral.

Dark green crystals of torbernite with vitreous lustre. Specimen from France.

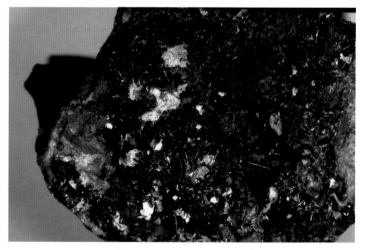

TURQUOISE

Group	Phosphates
Composition	$CuAl_6(PO_4)_4(OH)_8.4H_2O$
Crystal System	Triclinic
Habit	Crystals rare, prismatic; concretionary, massive, crusts, stalactitic
Cleavage	Perfect
Fracture	Conchoidal
Specific Gravity	2.6 to 2.8
Hardness	5 to 6
Colour	Bluish-white
Streak	White, pale green
Lustre	Vitreous

Turquoise occurs where igneous and sedimentary rocks have been altered by hydration. Acidic waters seeping into aluminium- and copper-rich rocks lead to the formation of turquoise. It is relatively hard, and with its bright blue colour turquoise has been used for thousands of years as a gemstone and decorative material. It has now been superseded to some extent by synthetic materials.

A crust of pale-blue turquoise on grey quartz, from Cornwall, England.

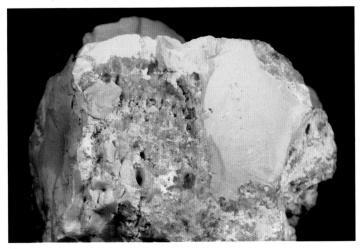

WAVELLITE

Group	Phosphates
Composition	$Al_3(PO_4)_2(OH,F)_3 \cdot 5H_2O$
Crystal System	Orthorhombic
Habit	Crystals acicular, prismatic; spherical, radiating
Cleavage	Perfect
Fracture	Uneven or subconchoidal
Specific Gravity	2.4
Hardness	3½ to 4
Colour	Greenish, white, yellowish, brown
Streak	White
Lustre	Resinous, pearly or vitreous

Wavellite commonly occurs on the surfaces of joints cutting through metamorphic and igneous rocks and in phosphate-rich sedimentary rocks in association with minerals such as quartz, muscovite mica and limonite. It is a secondary mineral, forming mainly in rocks rich in aluminium. Wavellite is readily identified by its spherical habit, with a radiating internal structure.

Rounded, greenish wavellite on rock surface. The broken specimens show the concentric internal layering. Specimen from Devon, England.

APATITE

Group	Phosphates
Composition	$Ca_5(PO_4)_3(F,Cl,OH)$
Crystal System	Trigonal/hexagonal
Habit	Crystals tabular, prismatic; granular, massive
Cleavage	Poor
Fracture	Uneven or conchoidal
Specific Gravity	3.1 to 3.2
Hardness	5
Colour	Green, grey, purple, colourless, white, yellowish
Streak	White
Lustre	Resinous or vitreous

Apatite occurs in many different rocks, including lavas and some sedimentary and metamorphic rocks. It also forms in phosphatic deposits of organic origin. This mineral defines point five on the hardness scale. When present in sufficient amounts, apatite is mined for fertiliser and as a source of phosphorus. It is sometimes used as a gemstone and a pigment in paint.

A hexagonal prismatic crystal of grey and brown apatite. Specimen from Brazil.

ARSENATES

Arsenates are minerals which contain metals combined with the arsenate radical (AsO_4).

ADAMITE

Group	Arsenates
Composition	$Zn_2AsO_4(OH)$
Crystal System	Orthorhombic
Habit	Crystals tabular; spheroidal, massive
Cleavage	Good
Fracture	Uneven or subconchoidal
Specific Gravity	4.3 to 4.5
Hardness	3½
Colour	Green, yellow
Streak	White
Lustre	Vitreous

Adamite occurs where mineral veins containing zinc and arsenic are altered by oxidation through weathering. It is found with quartz, calcite, malachite, limonite, oxides of manganese, hemimorphite, olivenite and smithsonite. Depending on its detailed chemistry and any impurities present, adamite may be fluorescent in ultraviolet light.

Masses of small greenish, vitreous adamite crystals on limonite-rich rock. Specimen from Mexico.

Mass of small acicular crystals of erythrite showing characteristic pinkish-purple colour and vitreous lustre. Specimen from Russia.

ERYTHRITE

Group	Arsenates
Composition	$Co_3(AsO_4)_2 \cdot 8H_2O$
Crystal System	Monoclinic
Habit	Crystals acicular, prismatic, bladed; earthy
Cleavage	Perfect
Fracture	Uneven
Specific Gravity	3.2
Hardness	1½ to 2½
Colour	Pink, purple
Streak	Purple, pinkish
Lustre	Pearly or vitreous

Erythrite occurs in areas where veins containing cobalt have been altered by oxidation and weathering. Its presence is often a good guide to prospectors searching for cobalt and silver. Erythrite sometimes occurs as a very thin purple or pink coating (bloom) on arsenic and cobalt minerals. It can be found in association with adamite, cobaltite, skutterudite and scorodite.

Greenish barrel-shaped mimetite crystals from Cumbria, England.

MIMETITE

Group	Arsenates
Composition	$Pb_5(AsO_4)_3Cl$
Crystal System	Monoclinic
Habit	Crystals prismatic, barrel-shaped; granular, botryoidal, reniform
Cleavage	None
Fracture	Uneven or subconchoidal
Specific Gravity	7.0 to 7.3
Hardness	3½ to 4
Colour	Brown, yellow, orange, green
Streak	White
Lustre	Vitreous or resinous

Mimetite occurs in the zones of lead-bearing mineral veins that have been altered by oxidation. It is found with galena, vanadinite, pyromorphite, arsenopyrite, quartz and hemimorphite. The small crystals, like those of pyromorphite, are frequently barrel-shaped, and this variety is called campylite. Because it contains arsenic, a strong smell of garlic is produced if mimetite is placed in a flame.

VANADATES

Vanadate minerals are composed of metals combined with the vanadate radical (VO_4) or (VO_3).

VANADINITE

Group	Vanadates
Composition	$Pb_5(VO_4)_3Cl$
Crystal System	Trigonal/hexagonal
Habit	Crystals prismatic; crusts, radiating, fibrous
Cleavage	None
Fracture	Uneven or conchoidal
Specific Gravity	6.9
Hardness	3
Colour	Orange, red, yellow, brown, white
Streak	Yellowish, white
Lustre	Subadamantine or resinous

Vanadinite occurs where mineral veins containing lead ores have been altered by oxidation. It is associated with galena, mimetite, wulfenite, pyromorphite, cerussite, barite and quartz. Though vanadinite is not a common mineral, it is popular with collectors. It is a source of vanadium, but is not mined for this purpose. Most economically used vanadium is produced as a by-product from other minerals, mainly in Russia and South Africa. Vanadium is used as an alloy with steel.

Prismatic, hexagonal crystals of vanadinite showing subadamantine lustre and typical brownish-red colouring. Specimen from Mexico.

Massive and crystalline descloizite with a greasy lustre and typical brownish-red colouring. Specimen from Namibia.

DESCLOIZITE

Group	Vanadates
Composition	$Pb(Zn,Cu)(VO_4)(OH)$
Crystal System	Orthorhombic
Habit	Crystals prismatic, pyramidal, tabular; botryoidal, crusts, mammilated
Cleavage	None
Fracture	Conchoidal or uneven
Specific Gravity	6.2
Hardness	3 to 3½
Colour	Brown, blackish, red, orange
Streak	Orange, reddish-brown
Lustre	Vitreous or greasy

Descloizite occurs where lead-bearing minerals have been altered by weathering and oxidation. It is found with a variety of hydrothermal and secondary minerals including vanadinite, pyromorphite, quartz, galena, cerussite, wulfenite and mimetite. Descloizite was, at one time, mined in northern Namibia for its vanadium content, but the deposits there are now exhausted.

SILICATES

Silicate minerals are composed of metals combined with the silicate molecule (SiO_4). Many silicates have very complex, lengthy chemical formulae. Silicates are very important as rock-forming minerals.

OLIVINE

Group	Silicates
Composition	Mg_2SiO_4–Fe_2SiO_4
Crystal System	Orthorhombic
Habit	Crystals tabular; granular, massive
Cleavage	Poor
Fracture	Uneven or conchoidal
Specific Gravity	3.3 to 4.3
Hardness	6½ to 7
Colour	Green, brownish, yellowish
Streak	Colourless
Lustre	Vitreous

Olivine occurs in basic and ultrabasic igneous rocks, especially basalt and peridotite. The ultrabasic rock dunite is composed almost entirely of this mineral. Olivine is also present in some types of marble, giving the rock a greenish tinge. As can be seen from the varying chemical composition of olivine, this is actually a series of minerals. The magnesium-rich variety is forsterite and the iron-rich end-member is fayalite. Olivine is one of the very first minerals to crystallise as lava or magma solidifies. It has been found in meteorites and rock specimens from the moon.

A massive specimen of olivine, variety peridot, showing typical green colouring and vitreous lustre. Specimen from Gran Canaria.

Grossular garnet, pinkish-red crystals in rock matrix.

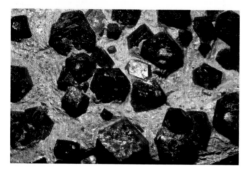

Almandine garnet, dark crystals in schist. Specimen from Norway.

GARNET

Group	Silicates
Composition	$Mg_3Al_2(SiO_4)_3$ $Fe_3Al_2(SiO_4)_3$ $Ca_3Al_2(SiO_4)_3$
Crystal System	Cubic
Habit	Crystals dodecahedral; granular, massive
Cleavage	None
Fracture	Uneven or conchoidal
Specific Gravity	3.4 to 4.3
Hardness	6½ to 7½
Colour	Red, brown, black, green, yellowish
Streak	White
Lustre	Vitreous

Garnet is the name for a group of minerals, the three main ones being magnesium-rich pyrope, iron-rich almandine and calcium-rich grossular. Almandine and grossular are found in metamorphic rocks such as schist and marble, whereas pyrope tends to occur in ultrabasic igneous rocks like peridotite. Many forms of garnet, because of their hardness and colour, are used as gemstones.

TOPAZ

Group	Silicates
Composition	$Al_2SiO_4(F,OH)_2$
Crystal System	Orthorhombic
Habit	Crystals prismatic, pyramidal; columnar, massive
Cleavage	Perfect
Fracture	Uneven or subconchoidal
Specific Gravity	3.5 to 3.6
Hardness	8
Colour	Colourless, grey, white, green, blue, purple, yellow, brown, orange, reddish
Streak	Colourless
Lustre	Vitreous

Topaz forms in very coarse-grained igneous rocks called pegmatites, in which crystals of great size can develop. Topaz crystals over 100kg in weight have been found. It also occurs as fine crystals in veins and joints in granites. For thousands of years topaz has been used as a gemstone because of its colours and hardness, defining point eight on the hardness scale.

Prismatic topaz crystals showing vitreous lustre. Specimen from Brazil.

ZIRCON

Group	Silicates
Composition	$ZrSiO_4$
Crystal System	Tetragonal
Habit	Crystals prismatic, pyramidal; fibrous, granular
Cleavage	Imperfect
Fracture	Conchoidal or uneven
Specific Gravity	4.6 to 4.7
Hardness	7½
Colour	Colourless, brown, grey, green, yellow, red
Streak	White
Lustre	Vitreous

Zircon occurs in a variety of igneous and metamorphic rocks, and is also well known in detrital sediments. Because of its hardness, zircon resists weathering and erosion. It may be carried from its original source before being deposited in alluvial sands, as in Western Australia. Zircon is an ore of zirconium and is mined as such in the USA, Sri Lanka and Australia. Some of the colour varieties are used as gemstones.

Brown vitreous zircon crystals from North Carolina, USA.

ANDALUSITE

Group	Silicates
Composition	Al_2SiO_5
Crystal System	Orthorhombic
Habit	Crystals prismatic; columnar, massive, fibrous
Cleavage	Distinct
Fracture	Uneven or subconchoidal
Specific Gravity	3.1
Hardness	6½ to 7½
Colour	Brown, grey, white, green
Streak	Colourless
Lustre	Vitreous

Andalusite forms in many acid igneous rocks, including granite and its very coarse-grained equivalent, pegmatite. In metamorphic rocks, including hornfels, it is found with corundum, cordierite and kyanite. Chiastolite is a variety of andalusite that has a cross-shaped section. This is found in spotted hornfels rock.

Pale, prismatic andalusite crystals in hornfels, from Brittany, France.

KYANITE

Group	Silicates
Composition	Al_2SiO_5
Crystal System	Triclinic
Habit	Crystals bladed; massive, fibrous
Cleavage	Perfect
Fracture	Uneven
Specific Gravity	3.5 to 3.7
Hardness	4 to 7½
Colour	Blue, white, yellow, green, grey
Streak	Colourless
Lustre	Vitreous or pearly

Kyanite occurs in metamorphic rocks, mainly the medium-grade rock schist and the high-grade rock gneiss. It also forms in quartz veins and pegmatites. Associated minerals include andalusite, staurolite, corundum, talc, hornblende and sillimanite. Kyanite has been found in detrital sands derived from metamorphic rocks. The hardness of kyanite varies according to which crystal face is tested and in which direction the crystal is scratched.

A mass of bladed, blue kyanite crystals from Brazil.

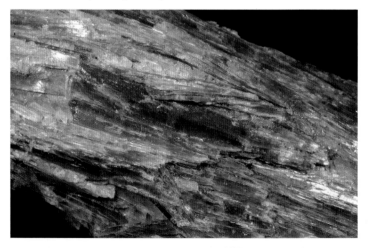

EPIDOTE

Group	Silicates
Composition	$Ca_2(Al,Fe)_3(SiO_4)_3(OH)$
Crystal System	Monoclinic
Habit	Crystals prismatic, acicular; granular, massive, fibrous
Cleavage	Perfect
Fracture	Uneven
Specific Gravity	3.3 to 3.5
Hardness	6 to 7
Colour	Green, brown, black, yellowish-green
Streak	Greyish, colourless
Lustre	Vitreous

Epidote occurs in metamorphic rocks such as schist and marble. It can also form by the hydrothermal alteration of a variety of silicate minerals such as feldspar, garnet, mica, amphibole and pyroxene. Epidote is found in some igneous rocks, including gabbro and amphibolite. This mineral often has striated crystal faces.

Crystals of dark, vitreous epidote, with prismatic habit. Specimen from Quebec, Canada.

ZOISITE

Group	Silicates
Composition	$Ca_2Al_3(SiO_4)_3(OH)$
Crystal System	Orthorhombic
Habit	Crystals prismatic; columnar, massive
Cleavage	Perfect
Fracture	Conchoidal or uneven
Specific Gravity	3.5
Hardness	6½ to 7
Colour	Brown, white, grey, greenish, pink, blue, purple
Streak	Colourless
Lustre	Vitreous

Zoisite forms in igneous pegmatites and in metamorphic rocks, such as gneiss and eclogite. It has also been found in hydrothermal veins, with a variety of sulphides, such as pyrite and galena. The prismatic crystals often have striations on their faces. Some varieties of zoisite are used as gemstones and colour varieties are named, the pink form being called thulite and the purple or blue variety tanzanite.

Granular, pinkish zoisite, variety thulite. Specimen from Tanzania.

HEMIMORPHITE

Group	Silicates
Composition	$Zn_4Si_2O_7(OH)_2.H_2O$
Crystal System	Orthorhombic
Habit	Crystals tabular, acicular; botryoidal, massive, fibrous, granular
Cleavage	Perfect
Fracture	Conchoidal or uneven
Specific Gravity	3.4 to 3.5
Hardness	4½ to 5
Colour	White, colourless, brown, green, blue, yellow, grey
Streak	Colourless
Lustre	Vitreous or silky

Hemimorphite occurs in regions where veins containing zinc minerals, such as sphalerite, have been altered. It is found with calcite, quartz, galena, sphalerite, smithsonite, cerussite and aurichalcite. The tabular crystals of hemimorphite are often different at each end, this hemimorphic structure giving the mineral its name.

Small clusters of minute blue hemimorphite crystals with vitreous lustre from Arizona, USA.

VESUVIANITE

Group	Silicates
Composition	$Ca_{10}Mg_2Al_4(SiO_4)_5(Si_2O_7)_2(OH)_4$
Crystal System	Tetragonal
Habit	Crystals pyramidal, prismatic; granular, massive, columnar
Cleavage	Poor
Fracture	Conchoidal or uneven
Specific Gravity	3.3 to 3.4
Hardness	6 to 7
Colour	Brown, green, purple, red, yellow, white
Streak	White
Lustre	Vitreous or resinous

Vesuvianite is found in zones where limestones have been affected by heat from igneous rock during contact metamorphism. It also forms in intermediate igneous rocks such as syenite. Vesuvianite has been used as a gemstone and colour varieties are named, cyprine being the blue-coloured form and californite the yellow or white variety. Vesuvianite was previously known as idocrase.

Prismatic crystals of brownish and greenish vesuvianite with vitreous lustre.

BERYL

Group	Silicates
Composition	$Be_3Al_2Si_6O_{18}$
Crystal System	Trigonal/ hexagonal
Habit	Crystals prismatic, pyramidal; columnar, massive
Cleavage	Poor
Fracture	Conchoidal or uneven
Specific Gravity	2.6 to 2.9
Hardness	7½ to 8
Colour	Green, pink, blue, yellow, white, colourless
Streak	White
Lustre	Vitreous

Beryl occurs in acid igneous rocks including granite and pegmatite. Huge crystals over 5 metres long are sometimes found in pegmatites. One crystal 18 metres long was discovered in Madagascar. The various colour varieties are used as gemstones and have individual names. Green beryl is emerald (p.75); yellow beryl is called heliodor; blue-coloured beryl is aquamarine and pink-coloured beryl is morganite. Beryl is the chief ore of beryllium.

A hexagonal prismatic beryl crystal with blue and pink colouring with a smaller crystal. Specimen from Brazil.

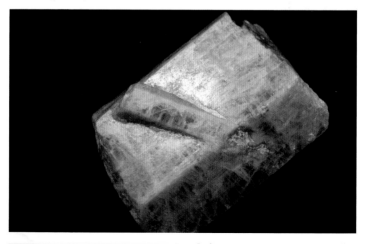

LEFT: Prismatic hexagonal crystals of tourmaline with typical vitreous lustre. Specimen from Afghanistan. RIGHT: Pink tourmaline crystals, variety rubellite, with quartz. Specimen from Cornwall, England.

TOURMALINE

Group	Silicates
Composition	$Na(Mg,Fe,Li,Mn,Al)_3Al_6(BO_3)_3Si_6(OH,F)_4$
Crystal System	Trigonal/hexagonal
Habit	Crystals prismatic; massive
Cleavage	Indistinct
Fracture	Conchoidal or uneven
Specific Gravity	3.0 to 3.2
Hardness	7 to 7½
Colour	Purple, green, black, brown, yellow, pink
Streak	Colourless
Lustre	Vitreous

Tourmaline forms in acid igneous rocks, especially granites, as an accessory mineral. It can also be found in schist and marble and in detrital sedimentary rocks. There are many named colour varieties of tourmaline which are used as gemstones; rubellite is pink tourmaline, schorl is black in colour and dravite is brown. Often they are different colours at each end, usually pink and green. As with beryl, tourmaline crystals can be of great size.

DIOPTASE

Group	Silicates
Composition	$CuSiO_2(OH)_2$
Crystal System	Trigonal/hexagonal
Habit	Crystals prismatic, rhombohedral; massive
Cleavage	Perfect
Fracture	Conchoidal or uneven
Specific Gravity	3.3
Hardness	5
Colour	Green
Streak	Pale greenish-blue
Lustre	Vitreous

Dioptase is a copper silicate mineral and occurs where copper minerals have been altered, often in arid regions. Associated minerals include quartz, calcite, chrysocolla and cerussite. Dioptase has been mistaken for emerald (green beryl), but can readily be distinguished, as it is less hard, at 5, than beryl (7 to 8). Also, dioptase tends to be darker green.

Prismatic dioptase crystals with vitreous lustre. Specimen from south-west Africa.

Prismatic, twinned augite crystals with characteristic dark colour and vitreous lustre. Specimen from Kenya.

AUGITE

Group	Silicates (pyroxene)
Composition	$(Ca,Na)(Mg,Fe,Al)(Si,Al)_2O_6$
Crystal System	Monoclinic
Habit	Crystals prismatic; granular, massive
Cleavage	Good
Fracture	Conchoidal or uneven
Specific Gravity	3.2 to 3.5
Hardness	5½ to 6
Colour	Black, dark green, brown
Streak	Grey-green
Lustre	Vitreous or dull

Augite is a common mineral in basic and ultrabasic igneous rocks. In basalt, augite may make up over 40% of the rock and gives the rock its dark colouring. Its crystals are usually rather squat. Augite is a member of the pyroxene group of minerals and can be confused with amphibole group minerals such as hornblende. The latter tends to form more elongated crystals, but the cleavage of the two is a good way to tell them apart. Augite cleaves along two planes virtually at right angles to each other, whereas hornblende cleaves at 60° and 120°.

SPODUMENE

Group	Silicates (pyroxene)
Composition	$LiAlSi_2O_6$
Crystal System	Monoclinic
Habit	Crystals prismatic; massive
Cleavage	Perfect
Fracture	Uneven
Specific Gravity	3.0 to 3.2
Hardness	6½ to 7½
Colour	Colourless, white, pink, lilac, yellow, green, grey
Streak	White
Lustre	Vitreous or dull

Spodumene occurs in pegmatites, where crystals can grow to great size, with beryl, tourmaline and quartz. It is sometimes used as a gemstone and different colour varieties are named. Lilac or pink-coloured spodumene is called kunzite, the colouring being due to small amounts of manganese. Hiddenite is green spodumene, coloured by chromium. Spodumene is an important source of lithium, and is mined in Australia, the USA, Russia and China. Lithium is of very low specific gravity and alloys with other metals are therefore very light, especially the alloy with aluminium used in aircraft. Spodumene is a member of the pyroxene group.

Massive pale-coloured spodumene showing cleavage. Specimen from Finland.

JADEITE

Group	Silicates (pyroxene)
Composition	$Na(Al,Fe)Si_2O_6$
Crystal System	Monoclinic
Habit	Crystals prismatic; granular, massive
Cleavage	Good
Fracture	Splintery
Specific Gravity	3.2
Hardness	6 to 7
Colour	Green, grey, white, brown, yellow, purplish
Streak	Colourless
Lustre	Vitreous or greasy

Jadeite is known from serpentinite, a highly altered ultrabasic rock, and it occurs in schist. It is valued as the ornamental stone jade. Another mineral, nephrite, is also used for this purpose and similarly known as jade. Nephrite is a form of actinolite, an amphibole mineral, whereas jadeite is a member of the pyroxene group. Despite their considerable hardness, these two minerals have been carved into intricate ornamental pieces for thousands of years.

Massive pale green and purple jadeite.

Prismatic, black hornblende cleavage fragment. Specimen from Czech Republic.

HORNBLENDE

Group	Silicates (amphibole)
Composition	$Ca_2(Mg,Fe)_4Al(Si_7Al)O_{22}(OH,F)_2$
Crystal System	Monoclinic
Habit	Crystals prismatic, bladed; columnar, fibrous, massive
Cleavage	Perfect
Fracture	Uneven
Specific Gravity	3.3 to 3.4
Hardness	5 to 6
Colour	Dark green, black, brown
Streak	Grey
Lustre	Vitreous

Hornblende is a common mineral in igneous rocks, especially those of acid and intermediate composition. The rock amphibolite, a highly altered metamorphic rock, contains a high percentage of this mineral. Hornblende belongs to the amphibole group of silicate minerals. These are often dark greenish minerals which can be mistaken for members of the pyroxene group. However, if the cleavage of hornblende (or other amphiboles) is considered, it will be seen that the cleavage planes intersect at almost exactly 60° and 120°, as opposed to the 90° of the pyroxenes.

Massive riebeckite showing fibrous structure and silky lustre. Specimen from Brazil.

RIEBECKITE

Group	Silicates (amphibole)
Composition	$Na_2(Fe,Mg)_3Fe_2Si_8O_{22}(OH)_2$
Crystal System	Monoclinic
Habit	Crystals prismatic; fibrous, massive
Cleavage	Perfect
Fracture	Uneven
Specific Gravity	3.3 to 3.4
Hardness	5
Colour	Black, dark grey, blue, brown
Streak	Bluish-grey
Lustre	Vitreous or silky

Riebeckite occurs in granites, pegmatites and schists. It is a member of the amphibole group. The form which occurs in a fibrous habit, crocidolite, used to have an industrial use as asbestos. The many applications of this material in fireproofing and insulation have now been banned because of the health risks from the minute fibres, which can be carcinogenic if inhaled or ingested. A variety of riebeckite with brown shimmering fibres is called 'tiger's eye' and is used ornamentally.

Small greenish acicular prismatic crystals of actinolite set in schist. Specimen from Scotland.

ACTINOLITE

Group	Silicates (amphibole)
Composition	$Ca_2(Mg,Fe)_5Si_8O_{22}(OH)_2$
Crystal System	Monoclinic
Habit	Crystals prismatic, acicular, bladed; granular, fibrous, massive
Cleavage	Good
Fracture	Subconchoidal or uneven
Specific Gravity	3.0 to 3.4
Hardness	5 to 6
Colour	Grey, greenish, blackish-green
Streak	White
Lustre	Vitreous

Actinolite forms in high-grade regionally metamorphosed rocks, one of which is called amphibolite because of the abundance of amphibole minerals. It can also be found in schist. A variety of actinolite called nephrite is used ornamentally as 'jade'. Actinolite commonly has a fibrous habit and then is a form of asbestos.

TALC

Group	Silicates
Composition	$Mg_3Si_4O_{10}(OH)_2$
Crystal System	Monoclinic or triclinic
Habit	Crystals rare, tabular; massive, fibrous, foliated, compact
Cleavage	Perfect
Fracture	Uneven
Specific Gravity	2.6 to 2.8
Hardness	1
Colour	Grey, greenish, white, brown
Streak	White
Lustre	Pearly, greasy or dull

Talc occurs in metamorphic rocks through the alteration of olivine, amphibole and pyroxene minerals. It defines point one on the hardness scale and has a noticeably greasy feel. This property has given talc a wide range of industrial applications, as a lubricant and in the manufacture of cosmetics, paint, plastic, paper and ceramics. Steatite (soapstone) is talc-rich schist. This is easily carved and used decoratively.

Massive greyish talc showing many surface markings because it is so soft. Specimen from Shetland, Scotland.

CHRYSOCOLLA

Group	Silicates
Composition	$(Cu,Al)_2H_2Si_2O_5(OH)_4.nH_2O$
Crystal System	Monoclinic
Habit	Crystals acicular, radiating; botryoidal, massive, earthy
Cleavage	None
Fracture	Conchoidal or uneven
Specific Gravity	2.0 to 2.4
Hardness	2 to 4
Colour	Green, blue-green, brown, black
Streak	White
Lustre	Vitreous or earthy

Chrysocolla occurs with other copper minerals where copper ores have been altered by oxidation. Usually it forms as crusts and rounded botryoidal masses, and is associated with limonite, quartz, cuprite, malachite and azurite. Though it is soft and readily scratched with a coin or knife blade, chrysocolla is sometimes cut and polished as a minor gemstone, mainly because of its green and blue colouring.

Greenish-blue chrysocolla with smaller pale green malachite spheres. Specimen from Australia.

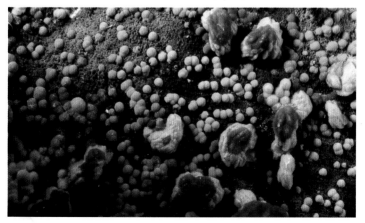

GLAUCONITE

Group	Silicates
Composition	$(K,Na)(Fe,Al,Mg)_2(Si,Al)_4O_{10}(OH)_2$
Crystal System	Monoclinic
Habit	Crystals extremely small, lath-shaped; tiny rounded granular aggregates
Cleavage	Perfect
Fracture	Uneven
Specific Gravity	2.4 to 2.9
Hardness	2
Colour	Dull green, yellowish, bluish
Streak	Pale green
Lustre	Dull or glistening

Glauconite occurs in sedimentary rocks deposited in the sea, especially on the continental shelf. It is diagnostic of a marine environment of formation for strata such as the Cretaceous Greensand, which is coloured by this mineral. It probably forms by the alteration of mica and other minerals during diagenesis.

Small greenish grains of glauconite in sandstone. Specimen from southern England.

Thin tabular twinned crystals of brownish muscovite mica from South America.

MUSCOVITE MICA

Group	Silicates
Composition	$KAl_2(Si_3Al)O_{10}(OH,F)_2$
Crystal System	Monoclinic
Habit	Crystals tabular; massive, scales, flakes
Cleavage	Perfect
Fracture	Uneven
Specific Gravity	2.8 to 2.9
Hardness	2½ to 4
Colour	White, colourless, grey, brownish
Streak	Colourless
Lustre	Vitreous, splendent or pearly

Muscovite is pale mica and it occurs in many igneous rocks, mainly those of acidic composition, mica being an essential mineral in granite. Metamorphic rocks such as schist contain much mica, which gives these rocks their glittery sheen on foliation surfaces. Crystals of muscovite may have a hexagonal outline. There are various mica minerals which differ in their detailed chemistry and colour. Cleavage fragments of muscovite are flexible and elastic and can be torn to leave jagged edges.

BIOTITE MICA

Group	Silicates
Composition	$K(Mg,Fe)_3(Al,Fe)Si_3O_{10}(OH,F)_2$
Crystal System	Monoclinic
Habit	Crystals prismatic, tabular; flaky, lamellar
Cleavage	Perfect
Fracture	Uneven
Specific Gravity	2.7 to 3.4
Hardness	2½ to 3
Colour	Black, brown, reddish, green
Streak	Colourless
Lustre	Vitreous or splendent

Biotite is dark mica. It occurs in similar rocks to muscovite and is an essential mineral in granite. The dark glittery specks in granite are usually flaky crystals of biotite. It also occurs in schist and gneiss; the darker bands in gneiss may contain biotite, as well as hornblende. The tabular crystals can have a hexagonal outline. Cleavage fragments are flexible. Biotite differs from muscovite in that it contains magnesium and iron.

Black biotite mica showing perfect cleavage and vitreous lustre. Specimen from Austria.

ORTHOCLASE FELDSPAR

Group	Silicates
Composition	$KAlSi_3O_8$
Crystal System	Monoclinic
Habit	Crystals tabular, prismatic; granular, lamellar, massive
Cleavage	Perfect
Fracture	Conchoidal or uneven
Specific Gravity	2.5 to 2.6
Hardness	6
Colour	White, pink, red, grey, yellowish, green
Streak	White
Lustre	Vitreous or pearly

Orthoclase feldspar (potassium feldspar) commonly occurs in many igneous rocks. It is an essential mineral in granite and some of these rocks have a pinkish colouring because of the presence of this mineral. Large crystals occur in pegmatite, and schist and gneiss contain orthoclase. Orthoclase feldspar can also be found in detrital sedimentary rocks such as arkose (feldspar-rich sandstone). Crystals of orthoclase are often twinned.

Pale prismatic orthoclase feldspar crystal with vitreous lustre. Specimen from Bulgaria.

Massive plagioclase feldspar with uneven fracture.

PLAGIOCLASE FELDSPAR

Group	Silicates
Composition	$(Na,Ca)Al_{1-2}Si_{3-2}O_8$
Crystal System	Triclinic
Habit	Crystals tabular; massive, granular, compact
Cleavage	Perfect
Fracture	Conchoidal or uneven
Specific Gravity	2.6 to 2.7
Hardness	6 to 6½
Colour	White, grey, blue, colourless
Streak	White
Lustre	Vitreous

Plagioclase feldspar is a group name for feldspars varying in composition from sodium aluminium silicate (albite) to calcium aluminium silicate (anorthite). There are a number of named varieties. These minerals occur in igneous rocks. The calcium-rich varieties tend to be in basic rocks such as gabbro and basalt; the sodium-rich forms are more common in intermediate and acid rocks. Plagioclase also occurs in metamorphic schists and gneisses. Crystals are often twinned in a complex way, with many sets of twins being joined. This is different twinning from the simple (two-fold) twinning of orthoclase feldspar.

Massive deep blue sodalite. Specimen from Brazil.

SODALITE

Group	Silicates
Composition	$Na_4Al_3Si_3O_{12}Cl$
Crystal System	Cubic
Habit	Crystals dodecahedral; granular, massive
Cleavage	Poor
Fracture	Uneven or conchoidal
Specific Gravity	2.1 to 2.4
Hardness	5½ to 6
Colour	Blue, greenish, yellow, white, reddish
Streak	Colourless
Lustre	Vitreous or greasy

Sodalite can be found in syenite, an igneous rock of intermediate composition, and can also form in veins and cavities, in association with calcite, fluorite, feldspars and nepheline. It has sometimes been found in meteorites. Sodalite is classified as a feldspathoid mineral. Though similar in general appearance to another silicate mineral, lazurite, sodalite rarely contains flecks of pyrite. Sodalite has been used ornamentally.

STILBITE

Group	Silicates
Composition	$NaCa_2Al_5Si_{13}O_{36}.14H_2O$
Crystal System	Monoclinic
Habit	Crystals rhombohedral, cruciform, bladed, radiating; globular
Cleavage	Perfect
Fracture	Uneven
Specific Gravity	2.1 to 2.2
Hardness	3½ to 4
Colour	White, yellowish, grey, pink, brown, red
Streak	White
Lustre	Vitreous or pearly

Stilbite is a zeolite mineral with water molecules in its chemical structure. It forms in vesicles in lavas with other zeolites, especially heulandite, and can also be found in hydrothermal veins, cavities in pegmatites and around hot springs. A characteristic feature of this mineral is its occurrence as sheaf-like aggregates of twinned crystals. It rarely occurs as single crystals.

Pale bladed crystals of stilbite on basaltic rock. Specimen from India.

GLOSSARY

Accessory mineral A mineral found in a rock, but having no bearing on the classification of that rock.

Acid rock An igneous rock which contains over 65% total silica, including over 10% quartz. These rocks are pale in colour and have a relatively low specific gravity.

Amphibole minerals Silicate minerals with a ferromagnesian composition and hydroxyl molecule(s).

Amygdale A cavity in lava infilled with minerals.

Basic rock An igneous rock with a silica content between 45% and 55%, including less than 10% quartz. Basic rocks are dark and have a relatively high specific gravity.

Batholith A very large igneous intrusion with an irregular shape. Batholiths can be many tens of miles in diameter.

Bedding plane A surface within a sedimentary rock. These rocks often break along bedding planes.

Cap rock An impermeable or non-porous layer of rock which prevents the upward movement of fluids, such as oil and gas, within the Earth's crust.

Compound Two or more elements combined chemically. Most minerals are compounds.

Country rock The local rock invaded by magma.

Diagenesis The various processes that turn loose sediment into a sedimentary rock.

Dyke A relatively thin sheet-like igneous intrusion which cuts across existing rock structures such as bedding planes.

Element Chemical material made of one type of atom.

Equigranular Having grains or crystals all the same size.

Fault A fracture in rocks, with relative displacement on each side.

Feldspathoid A mineral with similar composition and structure to the feldspars, but containing less silica.

Ferromagnesian Minerals composed of iron, magnesium and silica.

Fossil Any evidence of past life preserved in the rocks of the crust. Archaeological material is excluded.

Gangue Minerals which are of no economic value, found in hydrothermal ore veins.

Groundmass The matrix (mass) of a rock, which is often of a uniform grain size. Larger mineral crystals may be set in this groundmass.

Hydrothermal fluids Fluids in the Earth's crust, composed essentially of heated water containing many chemical elements.

Hydroxyl radical The molecule (OH).

Intermediate rock Igneous rock containing between 55% and 65% total silica.

Lava Molten igneous rock on the Earth's surface.

Magma Molten igneous rock below the Earth's surface.

Massive (1) Non-crystalline habit in minerals. (2) Sedimentary rocks with no obvious stratification are said to be massive.

Maturity Sedimentary rocks containing a high percentage of quartz are mature.

Phenocryst A relatively large crystal set in the finer-grained matrix of an igneous rock.

Placer deposit The alluvial accumulation of certain minerals which are preferentially deposited because of high specific gravity and/or hardness.

Plate tectonics The theory which considers the movement of the plates of the Earth's crust and the uppermost part of the mantle.

Pluton A large igneous intrusion.

Porphyritic An igneous rock texture composed of relatively large crystals (phenocrysts) set in a finer matrix.

Porphyroblast A relatively large crystal set in the matrix of a metamorphic rock.

Pyroclastic Describes fragmental material ejected from a volcano.

Pyroxene Ferromagnesian mineral without hydroxyl molecule(s).

Radical An atom or molecule that is active and readily combines with other atoms or molecules.

Replacement deposit The replacement of an original mineral deposit by another at a later stage in rock or mineral formation.

Rock texture The size, shape and orientation of the grains or crystals in a rock and the relationships between them.

Secondary mineral A mineral formed by the alteration of a previously formed mineral.

Semi-metal An element which has properties of both metals and non-metals.

Sill An intrusive sheet of igneous rock that follows existing structures.

Sorting A term referring to the grain size of a sedimentary rock. If the grains are all the same size, the rock is well sorted. A poorly sorted sediment has grains with a variety of sizes.

Strata Layers in sedimentary rocks, also called beds.

Striated Grooved. A striated crystal face has narrow parallel grooves.

Thrust fault Faults with fault planes at very low angles, where one mass of rock has been pushed up the fault plane over other rocks.

Ultrabasic rock Igneous rocks with less than 45% total silica content are ultrabasic in composition.

Unconformity A break in the geological succession. This represents a 'time gap' and at the unconformity part of the record of geological time is missing. Often the rocks below and above the unconformity are very different in type and structure.

Vesicular Having small gas bubble cavities.

Volatile A substance that readily vaporises.

Zeolites Hydrated alumino-silicate minerals, distinguished by their ability to lose and regain water molecules.

FURTHER READING

This short list of titles is a selection of the many available books on the subject. Further references are to be found in each of them.

Duff, D. and Holmes, A., 1993. *Principles of Physical Geology*, Nelson Thornes, Cheltenham.

Evans, A., 1997. *An Introduction to Economic Geology and its Environmental Impact*, Blackwell, Oxford.

Hamilton, W.R., Wooley, A.R. and Bishop, A.C., 1983. *Minerals, Rocks and Fossils*, Country Life, London.

Jerram, D., 2011. *Introducing Volcanology*, Dunedin, Edinburgh.

Keary, P., 2003. *Penguin Dictionary of Geology*, Penguin, London.

Park, G., 2010. *Introducing Geology*, Dunedin, Edinburgh.

Pellant, C., 1992. *Rocks and Minerals*, Dorling Kindersley, London.

Pellant, C. and H., 2005. *1000 Facts on Rocks and Minerals*, Miles Kelly, Essex.

Roberts, J.L., 1989. *Geological Structures*, Macmillan, London.

Roberts, W.L., Campbell, T.J. and Rapp, G.R., 1990. *Encyclopedia of Minerals*, Van Nostrand Reinhold, New York.

Smith, D.G. (ed), 1982. *The Cambridge Encyclopedia of Earth Sciences*, Cambridge University Press, Cambridge.

Stow, D., 2005. *Sedimentary Rocks in the Field*, Manson, London.

Watson, J., 1983. *Geology and Man*, Allen and Unwin, London.

WEBSITES

Listed below are a few useful sites. By searching the internet, a great deal of information about rocks and minerals can be found, though some sites are more reliable than others.

www.bgs.ac.uk	The British Geological Survey
www.usgs.gov	U.S. Geological Survey
www.geologist.demon.co.uk	Geologists' Association
www.nhm.ac.uk	The Natural History Museum
www.si.edu	The Smithsonian Museum
www.gemrock.net	Creetown Gem Rock Museum
www.richardtayler.co.uk	An excellent source of minerals for the collector

ACKNOWLEDGEMENTS

Over the years we have been assisted by many people and organisations, particularly with permission to photograph specimens in their collections. These include members of staff at the Geology Department, University of Keele, Ken Sedman (Cleveland Geology and Environmental Resources), Jim Nunney (Leeds City Museum) and Sid Weatherill (Hildoceras, Whitby). Richard Tayler has supplied us with many fine specimens. Emily and Martin Swan provided photographs of New Zealand and Australia. John Searby of Picturesk, Whitby, scanned many of our transparencies. We would also like to thank Simon Papps, who originally commissioned the book, Alice Ward for excellent editorial work and Susan McIntyre for the design. Many thanks to them all.

INDEX

Emmet Keogh: A crack gunman for the Irish Republic, he is fleeing from his past. Now, in a lawless Mexican town, he saves the life of a beautiful girl—and begins a perilous journey into the jaws of death . . .

Oliver van Horne: Beneath his priest's garb he carries a submachine gun. When he meets Keogh in a shower of bullets and blood, they plunge into a murderous mission . . .

Paul Janos: A huge, relentless man of shadowy origins, he will play all sides against one another. But now he's equal partners with Keogh and van Horne—a dangerous alliance that will end only with death . . .

Colonel Bonilla: The Military Governor is a cunning man with grand dreams. Now he offers his three unusual prisoners a proposition: the men can hang, or bring him the head of Tomás de la Plata . . .

Tomás de la Plata: The infamous revolutionary is now a fanatical outlaw and murderer. While he is alive, people will continue to suffer. But his death could be even worse for some . . .

BY THE SAME AUTHOR

JACK HIGGINS

(Previously Published as James Graham)

THE WRATH OF GOD

POCKET BOOKS

New York London Toronto Sydney Tokyo Singapore

The Wrath of God was originally published under the authorship of James Graham.

POCKET BOOKS, a division of Simon & Schuster Inc.
1230 Avenue of the Americas, New York, NY 10020

ISBN: 0-671-72454-1

First Pocket Books printing November 1990

10 9 8 7 6 5 4 3 2 1

POCKET and colophon are registered trademarks of Simon & Schuster Inc.

Cover illustration by Franco Accornero

Printed in the U.S.A.

For David Godfrey with thanks

MEXICO

1922

ONE

The Chief of Police usually managed to execute somebody round about noon on most days of the week, just to encourage the rest of the population, which gives a fair idea of how things were in that part of Mexico at the time.

The sound of the first ragged volley sent my hand down inside my coat in a kind of reflex action when I was halfway up the hill from the railway station. For most of the way I had managed to stay in the shade, but when I emerged into the Plaza Cívica, the sun caught me by the throat and squeezed hard, bringing sweat from every pore.

The executions were taking place in the courtyard of the police barracks and the gates stood wide open to give an uninterrupted view to anyone interested enough to watch, which on that occasion meant a couple of dozen Indians and mestizos. Not a bad audience considering the noonday heat and the frequency with which the performance was repeated.

At the rear of the small crowd, an automobile was

parked, a Mercedes roadster with the hood down, the entire vehicle coated with a layer of fine white dust from the dirt roads. An exotic item to find in a town like Bonito at that time. More surprising was the driver who was getting out just as I arrived, for he was a priest, although like no other priest I'd seen outside of Ireland—a great ox of a man in a shovel hat and faded cassock.

He ignored the rest of the audience, most of whom were surprised to see him there, produced a cigarillo from a fat leather case and searched for a match. I found one before he did, struck it and held it out for him.

He turned and looked at me sharply, giving me a sight of his face for the first time. A tangled graying beard, vivid blue eyes and the unmistakable furrow of an old bullet wound along the side of his skull just above the left eye. One of the lucky ones to survive the Revolution.

He took the light without a word and we stood side by side and watched as they marched three Indians across the courtyard from the jail and stood them against the wall. There were already half-a-dozen bodies on the ground and the wall was pitted with scars. The three men stood there impassively as a sergeant tied their hands behind their backs.

The priest said, "Does this happen often?"

He had spoken in Spanish, but with an accent that indicated that he was anything but Mexican.

I replied in English, "The Chief of Police says it's the only way he can keep down the numbers in the jail."

He glanced at me with a slight frown. "Irish?"